CHIEF SUSTAINABILITY OFFICERS AT WORK

HOW CSOs BUILD SUCCESSFUL SUSTAINABILITY AND ESG STRATEGIES

Chrissa Pagitsas

Apress®

Chief Sustainability Officers At Work: How CSOs Build Successful Sustainability and ESG Strategies

Chrissa Pagitsas
Washington, DC, USA

ISBN-13 (pbk): 978-1-4842-7865-9   ISBN-13 (electronic): 978-1-4842-7866-6
https://doi.org/10.1007/978-1-4842-7866-6

Managing Director, Apress Media LLC: Welmoed Spahr
Acquisitions Editor: Shivangi Ramachandran
Development Editor: Chris Nelson
Coordinating Editor: Mark Powers
Copyeditor: Kimberly Burton-Weisman

Cover designed by eStudioCalamar

Distributed to the book trade worldwide by Apress Media, LLC, 1 New York Plaza, New York, NY 10004, U.S.A. Phone 1-800-SPRINGER, fax (201) 348-4505, e-mail orders-ny@springer-sbm.com, or visit www.springeronline.com. Apress Media, LLC is a California LLC and the sole member (owner) is Springer Science + Business Media Finance Inc (SSBM Finance Inc). SSBM Finance Inc is a **Delaware** corporation.

For information on translations, please e-mail booktranslations@springernature.com; for reprint, paperback, or audio rights, please e-mail bookpermissions@springernature.com.

Apress titles may be purchased in bulk for academic, corporate, or promotional use. eBook versions and licenses are also available for most titles. For more information, reference our Print and eBook Bulk Sales web page at http://www.apress.com/bulk-sales.

Printed on acid-free paper

To my husband, Lee, and our sons, Cosmo and Felix

Contents

About the Author

Chrissa Pagitsas is a senior executive and advisor to global companies, philanthropies, and non-profits on the intersection of sustainability, environmental, social, governance (ESG) and business. She is a trusted board member and sought-after public speaker on how to catalyze business transformation through sustainability.

Chrissa was the driving force behind several market-transforming innovations in green financing, including issuing the first green commercial mortgage-backed security (CMBS) in the United States, creating a standardized energy audit for the commercial real estate market, and developing the US Environmental Protection Agency's ENERGY STAR 1 to 100 score for Multifamily Housing.

Recognized as a "Global Green Bond Champion" by the Climate Bonds Initiative, Chrissa served as the first vice president of ESG at Fannie Mae. She launched and led Fannie Mae's Green Financing Business, which under her leadership became the largest issuer of green bonds globally. Chrissa has developed sustainability strategies for large real estate asset managers, advised clients on renewable energy solutions, and delivered financial management software solutions for utility companies in the United States and Europe.

Chrissa holds an MBA from the Darden School of Business, University of Virginia, and a BA from Johns Hopkins University. She lives in Washington, DC, with her husband and sons.

Contact information and more available at www.chrissapagitsas.com.

Acknowledgments

The idea to write about leaders in sustainability became this book thanks to the support and encouragement of many family members and friends.

My deep thanks to my husband, Lee, who encouraged me throughout the writing process and crossed the finish line with me. Thank you to my sons, Cosmo and Felix, for many hugs at my desk and invitations to take a break and play. Thank you to my mother, Efi, who showed me what a determined woman could do in her life and career. Thank you to my sister, Elpi, whose ability to reframe challenges is inspirational. Thank you to Laurie for teaching me to think like an artist.

A heartfelt thank you to my friends from many parts of my life who cheered me on, including Kate Canale, Christine Davies, Gina Rajpal Felton, Erica Malkasian, James Shockley-Temple, and Katie Tower. Maria Barker, Christine Bohle Boyd, and Karyn Sper—thank you for helping me capture my story and why it was important to write this book.

An incredible group of colleagues helped build the green financing business and ESG strategy at Fannie Mae. I thank the Green and ESG teams, the multifamily business, corporate teams, and executive leadership for saying yes to new ideas, changing processes, and more over the last decade.

Thank you to the leaders in sustainability and ESG who agreed to be interviewed for this book. They not only shared their time but their ideas, passion, and vision for a more sustainable future. I am grateful that I can share their stories with you. An additional thank you to the people at each company, from assistants to marketing to legal, who helped finalize their CSO's chapter.

Thank you to Chris Nelson for his expert guidance on chapter development and narrative, Kim Burton-Weisman for her keen eye on copyedits, Mark Powers for strong project coordination, and Shivangi Ramachandran for the green light to write this book.

Preface

Behind a fire station and across the street from a jazz club in Cambridge, Massachusetts, there was a simple, brick apartment building. It had baseboard heating, a coin-operated laundry in the basement, and not much more. While growing up there in the 1980s, money was tight. My mother made sure my sister and I turned off the lights as we left a room and put on a sweater if we complained about the New England winter. Hot summer nights meant catching a cool breeze through open windows along with fire truck sirens and jazz music. Every dollar spent on the electric bill was a dollar lost toward our groceries and schooling.

Little could I have imagined that thirty years later, I would devote myself to helping millions of other families to live in more energy- and water-efficient apartment buildings, spend less on their utilities, and be comfortable in their homes in summer and winter. Figuring out *how* to do that would eventually lead to the writing of this book.

As a consultant to utility companies in the early 2000s, I began to observe a shift in the energy and commercial real estate industries. Real estate developers were adopting energy-efficient construction, installing solar panels, and reducing their energy demand from the traditional energy grid. This new intersection of business and the environment changed the direction of my career. I went from programming energy price forecasting systems to evaluating the viability of methane biodigesters in emerging markets and advising asset managers on the sustainability strategy of their commercial real estate portfolios.

In 2010, Fannie Mae hired me to start a "green initiative" for their multifamily business, returning me to the world of apartment buildings. My mission was to find a way to transform their core product, a commercial mortgage, to become "greener." The question was, how does one "green" a mortgage for apartment buildings? Not only did these green mortgages have to maintain profitability and low credit risk, but I was adamant that they also had to improve the quality of housing and lower the energy and water bills for low- and moderate-income families.

How did we begin to make apartment buildings more energy- and water-efficient? A partnership with the US Department of Housing and Urban Development led to the piloting of multiple green mortgage products and the issuance of the first commercial green bond. To integrate sustainability principles into the existing loan underwriting process, my team and I created a standardized energy audit report that an energy engineer could deliver quickly and a lender could understand and underwrite to. It also satisfied green bond investors' demand for rigor, consistency, and transparency. I partnered with peers across internal departments, originators at primary lenders, and stakeholders at industry associations to put new green mortgage products in the hands of apartment building owners. Together, these approaches catalyzed a transformation across the multifamily commercial mortgage value chain. As of year-end 2021, Fannie Mae had issued more than $100 billion in green bonds backed by green mortgages.

Over this same time, awareness and agreement had risen exponentially that business and civil society needed to act to combat climate change and social issues, thanks to the steadfast efforts of activists, non-governmental organizations, and the private sector. New frameworks such as the Sustainability Accounting Standards Board (SASB) and the Task Force on Climate-related Financial Disclosures (TCFD) had been established to treat environmental data with the same rigor as financial data. And investors in the United States and Europe began to ask questions about companies' environmental, social, and governance performance, or more commonly known as ESG. ESG was being adopted as a strategic approach to assess a company's purpose, risk management, and value, married with qualitative and quantitative data. Recognizing this shift in the market toward ESG, I raised the imperative to launch an enterprise ESG strategy with Fannie Mae leadership. In 2019, I became Fannie Mae's first ESG officer. I thank Fannie Mae's leadership for taking this leap of faith and continuing to drive the company's ESG transformation today.

As I built these strategies at Fannie Mae, I would turn to my peers leading sustainability, green financing, and ESG strategies for advice. We traded stories about what worked—and what didn't—when integrating sustainability into the core services and products of our respective global firms, how to secure the *yes* to pilot new technologies, and more. I realized that we most frequently talked about *how* to implement successful sustainability and ESG strategies, not *why*. We had all answered *why* we had to act to meet humanity's critical challenges, such as climate change and racial inequity. The fact that my peers held senior executive positions at global organizations meant that their executive leadership and boards had answered the *why* question, too.

The unanswered *how* questions we discussed were complex, and their potential answers differed depending on the industry. How do you accelerate the positive environmental sustainability and social impact at a company with billions in revenue and operating across multiple continents? How do you overcome internal inertia to launch new pilots or retool a process? How does the executive management team maintain focus on decades-long ESG goals when challenged by present-day quarterly financial targets? How do companies in different industries, from financial services to food and beverage, prioritize the issues material to their diverse stakeholders, from regulatory bodies to farmers? How do chief sustainability officers lead differently than their C-suite peers? What is the future of business when seen through a sustainability lens?

Continuing global social, economic, and health crises told me that, now more than ever, companies and their leadership would need and want to know the answers to these *how* questions about ESG and sustainability. So, in 2021 I sought the answers with the help of twenty-five executives from Fortune 500 companies and globally recognized brands in diverse industries. This book, *Chief Sustainability Officers At Work*, captures our conversations about the *how* of sustainability and ESG, and at times, we even revisited the *why*.

For simplicity, I use the term *chief sustainability officer*, or CSO, as an umbrella term for senior executives of sustainability, heads of ESG, and managing directors of sustainable finance. However, there is no such thing as a cookie-cutter CSO. Their professional experience spans procurement to marketing to think tanks, and they lead their company's sustainability strategies from London, Singapore, and Hong Kong to New York and Seattle. Not surprisingly, these executives are frequently entrusted with responsibilities beyond sustainability, such as enterprise strategy, stakeholder management, procurement, human resources, philanthropy, and more.

Yet, as I asked these executives how one leads a credible, scalable, successful sustainability and ESG strategy, I heard five themes repeated often: *priorities, patience, products, process,* and *partnerships*. Eventually, five complete answers coalesced around the themes and echoed my experience building and leading a green bond business and an enterprise ESG strategy.

- Know the *priorities*.
- Invest in *patience*.
- Integrate with *products*.
- Catalyze *process* and systems change.
- Forge new *partnerships*.

CSOs *know the priorities* of their external and internal stakeholders. With that knowledge, they focus their company's resources on them to drive environmental and social impact and financial value. Katherine Neebe of Duke Energy crystallizes the questions that every company must answer as it embarks on a sustainability journey: What issues matter to the company, and to which issues does the company matter? Even when the stakeholders are many, and their opinions are diametrically opposed, it's important to listen and identify how to authentically address them, as Michelle Edkins and Alexis Rosenblum at BlackRock point out.

CSOs must innovate to create the techniques, tools, and technologies to close the gap between what's possible today and their ambitious ESG goals for 2030 and beyond. But the level of innovation needed to close the gap requires CSOs and their companies to *invest in patience*. It may take years for the new solutions to be scalable, profitable, and impactful. Ezgi Barcenas and the executive management team at Anheuser-Busch InBev made that investment when they concurrently approved ambitious ESG goals and a new accelerator program to pilot and scale sustainable innovations.

How do CSOs make progress on ESG strategies in the meantime? By *integrating ESG and sustainability principles within new and existing products*. Virginie Helias at Procter & Gamble showed the power of this answer many times, whether developing a Head & Shoulders bottle made from recycled plastic or launching an education campaign for consumers to use cooler water in their washing machines and save energy.

CSOs *catalyze changes in processes and systems* to manage their risks and create positive outcomes. At PVH, Marissa (Pagnani) McGowan and her team changed the auditing process for thousands of garment factories, allowing PVH to better understand where they produce their clothing, support their workers, and increase environmental efficiency. At the macro level, Steve Waygood of Aviva Investors, raises the need to redefine the role of the G20, the United Nations, and the Organisation for Economic Co-operation and Development to better support humanity and the planet.

Last but not least, CSOs told me about the importance of *forging new partnerships*. Kara Hurst of Amazon credits public-private partnerships, such as The Climate Pledge and the Lowering Emissions by Accelerating Forest finance (LEAF) Coalition, for amplifying the call to action on climate change and implementing solutions to solve it.

Together these executives paint a picture of what it is to be a chief sustainability officer and what it takes to know *how* to address sustainability, not just *why*. These are timeless lessons for any leader, but especially for those seeking to forge a new sustainable and impactful way of doing business. Whether you are an experienced sustainability professional, joining a new ESG team, or a "non-sustainability" leader determining what sustainability means for your business line, these lessons will help you navigate opportunities and the road ahead.

Our planet is in desperate need of leaders like these CSOs. Once you know *how* to lead a successful sustainability and ESG strategy, I hope you join us.

—Chrissa Pagitsas
February 1, 2022
Washington, DC

Introduction

This book contains twenty-four in-depth interviews with chief sustainability officers (CSOs) and senior executives leading sustainability and environmental, social, and governance (ESG) at Fortune 500 firms and globally recognized brands. Each chapter captures rich conversations between the executive and the author on the intersection of sustainability and business.

These executives play a central role in their company's long-term growth strategies by stewarding the environmental and social impact of core products and services. The CSOs are navigating how to address climate change and social injustice while regulatory frameworks and investor expectations continue to shift. They give a glimpse into the future of sustainability and business and what motivates them daily to tackle our world's greatest challenges.

Whether you are a sustainability professional, member of the board of directors, or leader in your company's business lines or operations, this book will help you navigate this new sustainable way of doing business. The CSOs reveal how they successfully partner with their peers across the company and with public and private entities externally, a critical skill for any business leader.

In each chapter, CSOs share their companies' sustainability and ESG goals as of year-end 2021. Over time, the goals will likely change as they expand their company's ambitions, however, the lessons they share on how to lead enterprise strategies are enduring. Likewise, the technical language and the practice of sustainability and ESG are quickly evolving and broad. Throughout these chapters, the executives use different words for the same or similar concepts, such as *greenhouse gases* and *carbon emissions*. To maintain each interviewee's unique voice and stay true to the strategy they lead, these terms have not been standardized.

Who Should Read This Book?

- CSOs and ESG leaders and their teams
- C-level executives and board members

- Business leaders partnering with sustainability leaders and teams
- Sustainability and ESG consultants
- Students studying sustainability and business
- All others who interface with sustainability and ESG functions in their roles

What the Reader Will Learn

- How CSOs lead sustainability strategies at global billion-dollar companies to create positive environmental and social impact and financial value
- How the business landscape may change over the next five to ten years as a result of sustainability and ESG
- How to partner with internal peers and boards of directors to integrate sustainability into core business products and services
- How to navigate a moving regulatory landscape while creating meaningful and authentic partnerships with external stakeholders

Sample Questions

Each CSO was interviewed about their views on business and sustainability strategy, the broad field of sustainability, ESG, and sustainable finance, and their leadership approach and management style. The following questions provide an overview of the questions asked, although no two interviews were alike. Each conversation between the executive and the author generally lasted an hour and was edited for clarity and readability.

Strategy

- How do you accelerate a company's positive environmental and social impact and create value at a global company with billions in revenue?
- How do you overcome internal inertia to launch new pilots or retool a process?
- Who is ultimately responsible for the success of the sustainability strategy?

- How does the executive management team balance ESG goals with target dates of 2030 and 2040 with quarterly financial targets?

- How do companies in different industries, from financial services to food and beverage, prioritize issues material to their diverse stakeholders, from regulatory bodies to farmers?

Sustainability, ESG, and Sustainable Finance

- What are the primary pillars of your company's sustainability, ESG, or sustainable finance strategy?

- What ambitious goals have been set to address climate change and greenhouse gas scopes 1, 2, and 3 emissions?

- How do you use reporting frameworks such as the Task Force on Climate-related Financial Disclosures (TCFD) and the Sustainability Accounting Standards Board (SASB)?

- What role do the United Nations Sustainable Development Goals (UN SDGs) have in your company's strategy?

- How are financial services institutions helping clients in traditional energy and fossil fuel industries transition to a low-carbon future?

- What are the implications of the agreements made at the UN Conference of the Parties (COP) held in Glasgow in 2021?

- How do diversity, equity, and inclusion intersect with a sustainability strategy?

Leadership and Management

- How do chief sustainability officers lead differently than their C-suite peers?

- Where does the sustainability team sit within the organization, and why was it placed there?

- How has your role changed since you came into the position?

- What skills and experience do you look for when hiring people to join your team?

- What is your philosophy and approach to leadership as a senior executive?

- What were pivotal moments in your career that altered your approach to business?

Learn from these executives' hard-won experiences and become ready to lead your business into a more sustainable and equitable future.

Consumer Goods, Hospitality, and Entertainment

Beatriz Perez

Senior Vice President and Chief Communications, Sustainability and Strategic Partnerships Officer
The Coca-Cola Company

Beatriz "Bea" Perez is senior vice president and chief communications, sustainability and strategic partnerships officer for The Coca-Cola Company. In her role, Bea aligns a diverse portfolio of work against critical business objectives to support brands, communities, consumers, employees, customers, and partners worldwide.

Since 2011, Bea has served as The Coca-Cola Company's chief sustainability officer (CSO). She was the first to hold this role at the company. Bea develops and leads progress against comprehensive global environmental, social, and governance (ESG) goals focusing on circular economy for packaging materials, water stewardship, greenhouse gas reduction, and women's economic empowerment. She works to advance a global sustainability strategy designed to help grow the business while making a lasting, positive difference for consumers, communities, and the environment.

Bea is recognized for her business leadership as a member of the American Advertising Hall of Fame and was inducted into the PRWeek Hall of Fame in 2020. She has been recognized as a "Conservation Trailblazer" by The Trust for Public Land. She has been named one of the 50 Most Powerful Latinas by The Association of Latino Professionals for America.

© Chrissa Pagitsas 2022
C. Pagitsas, *Chief Sustainability Officers At Work*,
https://doi.org/10.1007/978-1-4842-7866-6_1

Chrissa Pagitsas: Sustainability is a very broad umbrella. What are the primary areas of sustainability on which Coca-Cola focuses?

Bea Perez: We focus on water stewardship, packaging waste, carbon reduction—or climate, reducing added sugar, and reshaping our portfolio for growth by doing things like promoting low and no-calorie beverage options and making smaller packages. More broadly, agriculture, women's empowerment, human rights, and diversity, equity, and inclusion are the top priorities. We look at the areas where we believe we can truly make a difference and align to our company's purpose. Also, our areas of focus are tied to our enterprise risk management. We consider these the business risks that we know we need to make sure we're helping to address or stay on top of.

We also conduct a lot of discussion with everyone from NGOs [non-governmental organizations] to community leaders to employees to investors, and consumer research to look at where we can credibly make a difference with The Coca-Cola Company beyond some of the obvious areas. We also rely on consumers and government partners to help guide us where we can make a difference in shaping policy.

Pagitsas: It seems that for Coca-Cola sustainability is not narrowly environmental sustainability.

Perez: Correct. I love that ESG is finally the popular word because it encompasses more of what we do. And I love that investors are paying attention to those areas.

Pagitsas: You're right. Investors are paying attention to sustainability and ESG in a way that they may not have been three to five years ago and certainly not fifteen years ago. Why do you think they're now starting to focus on it?

Perez: I'd say it was a different kind of investor who paid attention fifteen years ago. They were the long-term focus, social-impact investors. They have always been there. They saw ahead of many other investors. Many investors were previously looking more at the financial data and not necessarily connecting that a company does well, especially for a consumer brand, if the community wants it there and values what it has to offer them.

I'd say more investors have started recognizing that strong financial performance requires success with the community. Investors fall on a spectrum. Some of them may still view it as risk management or ask how we mitigate risks. Then there are others who are beginning to understand that this next generation of consumers expects brands and businesses to have a point of view on certain social issues, to be doing the right things for the environment, and to be mindful of the resources that they're using.

Today, we still have some room to improve in terms of how investors give credit or value to companies in terms of sustainability, but I'm glad to see that they're all moving in that direction.

Pagitsas: You've mentioned "risk" and "risk management." Why is a beverage company managing risk through environmental and social issues versus traditional operations such as distribution, bottling, and other operations?

Perez: We actually do both. Risk management is absolutely core to how we have always thought of our business, and it's reflected in our public statements. What's changed, I'd say in the last ten to fifteen years, is that environmental, social, and governance areas have started to come into the enterprise risk matrix. Even though my role was created in 2011 officially as a C-suite role, the company had focused on environmental issues such as water and agricultural ingredients in terms of enterprise risk for a long time. They are two large inputs to a beverage business.

Let's take water. Water has been a risk for us to manage for a very, very long time. I think there is this realization across the industry that there is no beverage business without this input. If we're not managing these high-risk areas for our business, then it won't really matter what we do in distribution or with other parts of our operation because we can't produce the product without water. When we brought environmental risks into that enterprise risk matrix, we started to make better decisions. We can be more efficient in the business. We saw that managing water risk allowed us to free up some of those inefficiencies to protect the business and reinvest in other places where we knew we still had more work to do.

Today we look at the full risk picture. We make sure this is a part of the routines of the business where enterprise risk is discussed. We also see sustainability issues in terms of growth opportunities and how we connect with the consumer and the community. We pull it into the marketing opportunity lens and think about how do we grow brands through the work that we're doing.

Pagitsas: Bea, it sounds like sustainability is integrated into multiple points of the business, from its growth strategy to its consumer strategy.

Perez: It is, and it continues to be a journey. I believe we can always do better and go further.

Pagitsas: You mentioned your role as chief sustainability officer was created in 2011. But you've been at Coca-Cola much longer. How did you end up in this role?

Perez: I've been at Coke for more than twenty-five years. So, a long time! And I spent most of my time in the business, whether field operations or in marketing. I also have done some short-term time at bottlers, getting to know the business. Working in the business is very much how we train and teach our people.

While I was the first official C-suite officer as the chief sustainability officer, I was not the first person to manage environmental issues. We had a legacy of strong technical and scientific experts. There had been a lot of work done around environmental issues, and our foundation was already working on different community efforts. However, the beauty of the approach of the chairman and CEO at the time, Muhtar Kent, who appointed me into this role, was that he gave me the time to learn. I asked him two important things. The first one was, "Why me? Why do you want me to do this when you know that I'm not the expert?"

Pagitsas: What did Muhtar say?

Perez: It was interesting because he said, "Well, it's you because we have a lot of scientists, we have a lot of people already doing the work. However, the work needs to be better integrated into and treated as the business." He decided that he wanted to pull someone out of the business that had already worked in operations and would know many of the operators, which I did. He also wanted someone who could come in and start to learn about what we've been doing but then build it into business planning processes and truly embed it into the company.

That was my first "aha!" moment about sustainability. I said, "You're okay that I don't know everything deeply about climate or water? This is about how we treat this as a business and how we integrate it into the business?" He said, "Yes. That's exactly right."

The second important question I asked him was how to approach this. He said, "Well, most people don't get the blessing of time. So, I'm going to give you six to eight months. I want you to go around the world, meet with all the people who are doing the work, come back over, and tell us what you think we need to keep doing. And, what should we get rid of? What should the plans look like? I want you to come in with your observations." So, I did just that. I went around the world to some places where we were known to have issues.

Pagitsas: So, Muhtar wanted you to come back with on-the-ground information about environmental issues?

Perez: That's right. For example, we had already had an issue in terms of water in Kerala, India, with the community protesting our water use. I wanted to go there and understand what had happened and why it was an issue. I remember that trip vividly. I remember going to the location where people had been protesting us back in 2000 when the water issue took place.

Pagitsas: Was there another "aha!" moment in Kerala?

Perez: The "aha!" was that it didn't matter if we could scientifically prove that we were not depleting the water resource in the community. It didn't matter because we still had big red trucks that contained water driving by communities who didn't have water. So, they looked at Coca-Cola to answer questions such as, "What are you doing to truly make a difference? How are you fixing the bigger problem?"

What Muhtar allowed me to do was to learn. What I learned is that we needed to lean in and speak to the experts. But it was also clear to me that we needed an integrated plan. We did need a business plan, and we needed short- and long-term plans. We needed not only the plan that we presented to the operators or the plan aligned with the bottlers or with the community, but we also needed an overarching strategic plan for the subplans. The plan had to answer how we were going to get there. What was the hard work that needed to be done to move in that direction?

From the visits to locations across the world, we were able to gather insights and develop a more strategic plan and align on the top priorities. We also started to have it set in people's objectives in the operations. It was already in certain objectives, but it needed to be in the operational objectives as well. While we were doing all of that, part of the plan was to get our top investors to understand why this was so critical to the growth of the company. That was the ten-year plan back in 2011 and 2012. I feel like we are still executing it today.

Pagitsas: From your experience in Kerala and your world listening tour, my takeaway is that you were discovering new information.

Perez: Yes, that's correct. I learned where we thought we were helping and maybe where we were not really helping, so we had to adjust the plan. I'd even say that in the early days of when we set the water goal, it was bold, ambitious, and honestly, we did not know how we were going to get there. When we first started, we had a one-size-fits-all kind of strategy. As you get into the water discovery, you realize that won't work. There are water quality issues in some places and water scarcity issues in others, for example.

After a year or two of starting to roll some things out in regard to water, we needed to shift. We needed to get deep into addressing the local community issue and what the community needed from us. For annual plans and board presentations, though, you roll up the projects up to the aggregate goal.

Pagitsas: What is Coca-Cola's sustainability strategy related to water today?

Perez: We just set our new 2030 Water Security Strategy. We strongly believe that greater action is needed to protect the water resources that we share with ecosystems and communities and to support those who are most vulnerable. Unfortunately, water scarcity and water quality challenges are growing in many parts of the world. Two billion people still live without access to safe drinking water and sanitation.

To meet this great challenge, we have developed a new, ambitious global water security strategy. It is to achieve water security for our business, communities, and nature where we operate, source ingredients, and touch people's lives by 2030.

To realize this vision, we have set three priority outcomes. One, reduce, reuse, recycle, and replenish the water we use in bottling plants and their products where it is needed most. Two, enhance community resilience through water and sanitation access and climate adaptation, focusing on women and girls. And three, improve watershed health where we operate and source ingredients while continuing to achieve 100 percent replenishment.

Together with our bottling partners and stakeholders, we believe we can make a big, positive difference at our bottling facilities and in the regions where our ingredients are grown.

Pagitsas: Due to the COVID-19 pandemic, 2020 was a year where many companies had to shift their whole business strategies, let alone their sustainability strategies, and continued shifting through 2021. You instead launched an ambitious water goal. Where else did you pivot due to COVID-19?

Perez: Coca-Cola has set a "world without waste" goal. Collecting packaging waste is very important to the goal—but everybody is in lockdown. How do you collect packaging waste? Well, we had to lean in and pivot into design with a new, more sustainable path with new materials. It taught us a lot about being agile. It is important because you can't control everything. While you can control a few things in normal circumstances, the world is not normal right now and facing lots of challenges, depending on what region you're in. What a time we're in as a society!

We cannot give up on the bigger strategic picture and overarching sustainability goals. We are able to try new ways or innovative solutions to get there. As something new comes in, do you add it in or not, and how do you add it? We've had many of those pivot moments.

Pagitsas: Do you anticipate other large changes for either Coca-Cola or the beverage industry as a result of COVID-19?

Perez: Well, I think we've all had to change. Let me start with that. For us, about half of our beverage business comes from the "away from home" channels. These are places where people congregate. As a business, we had to pivot because, as you can see in our 2020 earnings, we saw a significant drop in our business. This was because as soon as everything closed, particularly restaurants, hotels, and cinemas, you're not selling Coca-Cola and other beverages to people.

In February 2021, when we announced our full-year 2020 earnings, you saw the business begin to recover by the back half of 2020. This was because we had had to get more innovative. We had to help some of our local customers figure out how they could add a service to provide beverages for at-home delivery. How could we help them by providing discounts? It was an all-in partnership between the company with our customers and even within the whole beverage industry.

At the same time, we were making sure that we put people first. With business plummeting in early 2020, a lot of people would say, "Cut the things that aren't going to give you an immediate return." But I give credit to our chairman and CEO, James Quincey, for not doing that.

James and I had a very open discussion. I said, "What do you think we need to do in sustainability now to build for the future?" He said, "Lean in, figure out what we can do today, do not stop any work. If you run into barriers, come see me." The next person I spoke to was our CFO, John Murphy. I asked him the same question, and he said, "I think the communities might actually need our work more right now. Let's do what we can."

Our global foundation, which I also lead, leaned heavily into supporting everything from food programs to health and mental well-being programs where people needed the support most. We went to each community and asked, "What do you really need?" There were blood drives in certain countries with the American Red Cross, and there were Feeding America programs and Boys & Girls Clubs of America programs. It depended on what the community told us. I will tell you it was the most complex series of conversations because it challenged my team. Our local teams that work closely with the foundation picked up the phones and said to partners, "We are expanding the funding from the foundation. We want to know what your communities need." Our work inside the company and the separate work of the foundation led to a total of $100 million combined commitment from the company and the foundation to help COVID-19 relief.

Similarly, how do you roll out vaccine distribution in the communities that may not otherwise get them? The foundation did a grant with Project Last Mile, which we've used before for programs to get HIV, TB, and malaria medicines to local communities in partnership with the Global Fund, USAID [US Agency for International Development], and the Bill & Melinda Gates Foundation. The foundation tapped into that infrastructure and created a partnership in Africa to do a grant around vaccine distribution. Now, communities have more access and the means to say where they're struggling and where they need that additional support.

Pagitsas: In the same year as the COVID-19 pandemic, we saw, particularly in the United States, ground-shifting conversations on social justice and the importance of the Black Lives Matter movement. How does Coca-Cola engage with these issues?

Perez: The Coca-Cola Company believes we have a duty to strive for greater justice and equity within our own company and throughout society. Coca-Cola is headquartered in Atlanta, Georgia, which has a long history in the civil rights movement.

James Quincey chose to speak to employees first. He called me and said, "I want you to arrange a town hall." I asked, "Do you need anything?" He said,

"I know what I need to say," and you could tell he was very emotional. Even though I'm also his communications head, I can tell you why I had nothing to do with what he said. It touched me very deeply, and he wanted to do more.

At the same time, what was happening in the regulatory space was that there were US states that did not have hate-crime legislation. Georgia was one of those. As a result of that employee town hall, a group of employees said, "We want to march to the Georgia State Capitol, and we want our positions to be heard in support of hate-crime legislation." Coca-Cola had already written letters to the governor and the lieutenant governor stating, "This law needs to pass." It was inspiring because employees led their own march. We helped make sure they were safe. James demonstrated strong support all the way up at the leadership level and inspired employees to act on their own and feel that they had the company's support.

Now just recently, members of the Asian and Pacific Islander community were affected by violence in Atlanta. We held what we call a Stand As One town hall soon after to let employees know that we stand in solidarity with our Asian and Pacific Islander colleagues and communities. It was a forum for employees to have a dialogue and express how they felt and what they've been experiencing. I'd say that the pandemic allowed us to lean into the social issues that we've had points of view in before, but also to engage in a two-way dialogue and to hear from the employees directly.

Pagitsas: You said that James encouraged you to keep moving forward and not hit the brakes on the sustainability strategy and goals. But there must have been challenges to the strategy and goals, particularly in 2020.

Perez: That's where the pivoting came in again! I was scared to death that we would miss some of our goals because of the pandemic. The company set these goals ten years ago without any idea there would be a pandemic, of course. And, you're rarely the person who set the goals ten years ago, and you're the person in the same seat who has to deliver them.

Pagitsas: True. It's very rare.

Perez: Very rare. I remember when the pandemic started coming around the world, and we started to recognize this was going to be big. I thought, "We're supposed to meet our 5by20 goal by 2020." The 5by20 program is our company's strategy to help five million female entrepreneurs—from fruit farmers and artisans to recyclers and retailers. Empowering women is one of the priority areas for Coca-Cola because women invest much of their income into their local economies. When we help women, we help communities.

In 2019, we had about a million and a half entrepreneurs still to reach, and I thought, "How are we going to do this?" I kept thinking to myself, "I'm going to have to go out and publicly explain why we're not going to make any of these goals." But when I went to James and John, they said, "We will make these goals. This is what people need. This is what we're going to prioritize. You do what you need to do."

Pagitsas: You've got me on the edge of my seat. Did you meet your 5by20 goal? How did you do it?

Perez: Yes, we did! Big congratulations are due to the whole set of operations around the world. We had been working very closely in the last several years with our 5by20 plan. Every operating unit or business unit had very important goals that they had to deliver within their plan. We knew what the allocation was per unit to get to the five-million goal. Some knew that, during this time of the pandemic, they needed to pivot to over-deliver because it was more challenging for some of the other operating units to deliver on their plan allocation.

It made me so proud of our operators and our bottling partners. When we met with them earlier on in the year, I thought a lot of them were going to say to me, "We can't do it. We need more time." Instead, all of them leaned in and said, "We will get there. We're so close. We're going to get there. This is where we can focus." For me, that was just amazing to see that it was no longer just Bea Perez and the sustainability team who really wanted to get there. It was the whole operation. It was everybody.

Pagitsas: Bea, what I'm really struck by is the teamwork at Coca-Cola around sustainability. It seems like you have the whole organization behind you. Coca-Cola's sustainability goals are not Bea's responsibility. They are the whole Coca-Cola team's responsibility. Is it critical that everybody own sustainability?

Perez: Yes, absolutely.

Pagitsas: Where does Coca-Cola's sustainability journey begin?

Perez: When I first came into the role and was on my listening tour, I found we had over seventy-five publicly stated commitments. They were all over the place! Almost none of them were set by the CEO. They were set by a regional manager or by an advertising manager. A lot of them didn't have end dates, and they weren't totally clear on what we were doing. I put them all on a spreadsheet, and I gave them to the Coca-Cola archivist. I thought, "This is interesting. We have been doing so much."

I also met with other companies like Unilever and said, "What are you all doing in this space?" I found we weren't alone. Everyone was trying to work hard and do a lot, but there were very few things that were directly attached to the business. As you know, this is a for-profit business. I also found that people were almost embarrassed to say that there was self-interest in here for the business. For me, I needed people to see the difference between the foundation and the business. The foundation is where we did things for society that had zero self-interest for the business. For me, sustainability became what was most meaningful to the business.

Sustainability was where we asked questions like, "What risks are in the enterprise risk matrix but also allow you to grow? What did your consumers—the people purchasing the brands—want to see?" That's where we got clear. That was where we started to say, "There are going to be no more than ten global metrics related to sustainability." There can be some freedom in a framework for some local flexibility based on local needs. For sustainability, we needed to say, "Well, water is the number-one ingredient for the product and the number-one lifeblood for the community," and get crystal clear on what that meant.

Number two for us was climate, packaging, and refrigeration. What are you delivering beverages in? It's a package, or it's equipment. Using hydrofluorocarbon-free equipment, making sure we had energy management systems—all of that was combined in there.

Women are another primary focus for us. Why women? Well, in many markets, the majority of purchasers of our products are made by women. They also happen to run a lot of retail locations in a lot of countries. They're often the key decision point in the retailer. And there's a large chunk of farmers who are women.

Just start to get clear on what is important to your business, what's important to society, and where you can make a meaningful difference. Don't try to do seventy-five things. Try to do maybe ten or three. For me, when I focused on these issues, the strategy began to crystallize. So, through this listening tour and this focus in 2011, we evolved the framework.

Pagitsas: How has the framework evolved over the last ten-plus years?

Perez: In 2011, we called the framework "Me, We, World." It was interesting at the time because its focus was on me as an individual, what we can do better as a community, and what we can do for the world. What we found, though, is that it was very hard to explain to consumers. As we talked about water, women, and packaging waste, we said, "Well, how about if we still have the overarching framework to set the goals and measure the risk and make the difference internally, but externally we go to water, women, and waste." We just go to the top issues.

If you look at my original documents and archives, you'll see that we originally had well-being and weather as our issues. When I spoke to some of the thought leaders in the climate space, they said, "You cannot equate this to weather. People will misunderstand." Now, we are very specific about what we do. You still have the broader framework, and now we have it ladder to ESG.

To fast-forward, real integration happens once you have a clear framework aligned to the business, clear objectives with goals, with the dates in there. We then started to assure the data with EY, [a global consulting firm], starting in 2013. I was originally told by EY, "No, this is not our core capability." I went to our CFO at the time, Gary Fayard. I said, "How can we make this part of EY's core capability? Because I don't feel like we have credibility if we can't assure it through the financial firms. I can get a third party to do this, but how independent will they be?" The credit goes to Gary, who called up his contacts at EY and said, "We would like to add this to our business. Come on the journey with us." They understood that we were going on a journey with or without them, so it was good to have them with us.

Pagitsas: It looks like you brought more than Coca-Cola on this journey! How has the board been engaged on this sustainability journey?

Perez: I'm lucky that I've had a board committee this whole time that has evolved as we have. The board committee just recently changed its name from the Sustainability Committee to the ESG and Public Policy Committee. We did that in the last meeting, and so credit goes to the board of directors. In my committee, I'm lucky. I have the chairman of the board, James Quincey. I have the lead director, Maria Elena Lagomasino. I have the chair of the Talent and Compensation Committee, Dr. Helene D. Gayle.

Quite frankly, having the head of the Talent and Compensation Committee on our ESG Committee is because of the next wave of our journey. In 2015 or 2016, we started talking about putting these sustainability goals into short-term compensation for the executive competency pillars, which was done. If you think about it, the people who were initially setting a lot of the objectives and showing progress all have a stake in the numbers that we've met today, even if they're no longer with our company. The Talent and Compensation Committee has been great at seeing the opportunity more overtly. That's not done yet, but it's close to being done now that the pandemic is in a more manageable place.

I would encourage a company beginning its journey to start with whatever is important to the business and where it can make a meaningful difference to its stakeholders. Then start to drive it into annual planning and long-range planning. Don't forget to use accounting firms and to integrate this into compensation.

Pagitsas: Is there a moment that significantly influenced how you personally lead your direct team?

Perez: Yes, there were definitely moments. I'm going to start by saying that I feel like I won the lottery when I got to Coca-Cola. You know, I was an aspiring lawyer. But I dropped out of law school to go to Coca-Cola. I thought

I'd defer my law school education for a year, and I'd go back. I wasn't sure because I did not like law school, even though it was one of my childhood dreams. Once I got into the academics of it, I thought, "I do not want to do this." I was a little bit lost.

I go into the past to say that, for me, it's about persevering. I got this from my mother, who left us in 2018 for the other world. She is the one person who said to me, "Never give up if you really want something that your heart of all hearts thinks is right." I first applied to Coca-Cola when trying to figure out what I wanted to do in life. I had been volunteering at an organization called the US Hispanic Chamber of Commerce.

While I was working there, I applied to Coca-Cola and I got rejected. I was sad. I went home to my mother and said, "I don't think they want me." She laughed and said, "They don't even know who you are. Why do you think that they don't want you?" She said, "Why don't you help them understand why they should want you?"

Pagitsas: How did you eventually make your way into Coca-Cola?

Perez: It was interesting because I didn't give up. Ultimately, I got in in a very roundabout way. I found a nice creative way to work for one of the company's agencies. I asked the creative marketing agency to put me into the Coca-Cola headquarters as the account representative in Atlanta versus working in San Antonio, where the agency was based. I was very honest with them, and I said, "My dream is to ultimately work for the client, and so if I can help you all and also help myself, I want to do this."

My mother loved my creativity and the client, Coca-Cola, accepted me as their account manager. They also said, "You have two weeks to get here." I loaded whatever I had into my car and left. In the early days, I had an empty apartment that no one knew was empty. I literally slept in a sleeping bag on the floor. I had my clothes, and that's it. It was all about perseverance. That moment really shaped me. Remember, I was on the agency side within the client building, not an employee. But what I found was this amazing group of people. Once I told people what I was really interested in doing, they all tried to help me.

When I got to the Coca-Cola building, I was supposed to meet up with the person who had offered me to work at Coca-Cola as an account manager. But I got lost. Some lady stopped to help me. She happened to be this phenomenal lady, Ingrid Saunders Jones, who was the longest-standing chair and president of the Coke Foundation. She's the first person who saw me, and she said, "You look lost. How can I help you?"

I wanted to mention this because I found a company where the people are absolutely extraordinary. I think that they're all incredibly smart and talented and caring. Caring is important to me. At the same time, it's a company where you can do anything you want. You can work around the world. You can work in marketing. You can even work in our fitness center if you want to. There's so much opportunity. I say this because what shaped me was a couple of different key opportunities during my career at Coca-Cola.

Pagitsas: What were those opportunities or influential moments that shaped your career?

Perez: Once I went to the client side, I got on the brand management path. An individual named Gary Azar ran the Eastern region—one of the big bottling territories. He ran fifty percent of the United States as an operator. I saw him present one day, and I reached out and asked him a few questions. We ended up scheduling a lunch. We went to a drive-through at McDonald's, and he made me get my own drink from the Coke vending machine. He's still my mentor today, even though he is now retired from Coke after forty years. We still talk.

He was the one who gave me the opportunity of a lifetime to be demoted. He said, "If you really want to learn this business, shape your career, and know what matters, you have to get closer to the consumer. You have to get closer to the people buying our brands." He also said, "You're in too high of a position in brand. To learn our field jobs, you need to go lower. You can't go in laterally because you don't know the area."

I remember once again calling my mother while I'm sitting in Atlanta all by myself. I asked, "Do I take this demotion? What do I do?" Gary was pretty serious. He said, "It's a demotion. I'm cutting your salary because you should not get paid what other people get paid. That wouldn't be right." He was right, and I took it. I remember thinking, "Why am I doing this?" But I was so excited. I then told myself, "This is the right thing to do." The people I had been reporting to said, "Are you nuts? Your career is on an upward trajectory." But what Gary said to me really stuck. At that lunch, he told me the story of what he calls the "merry-go-round."

He said, "In corporate America, there's a merry-go-round. You can go around, you'll go up and down, and you'll make a difference. You'll feel good about what you've done, but you're only going to go so high." Remember, on a merry-go-round, the poles can only go so high. He said, "If you get off the merry-go-round, it's going to be scary because you might fail. Or you might hit lots of stumbling blocks. You might not be popular." He said, "But you're going to learn. And then one day, you might realize that you'll be the operator running the merry-go-round."

It was amazing. That's why I said yes to him. He's been the toughest person I've ever worked for, and I probably was rated the lowest working for him. But he was great. He really taught me the business. In particular, he taught me something he calls "the arithmetic of the business." He quizzed me all the time on the math. He said, "You've got to learn math in your head, not on calculators. You've got to know how to do conversions" because our business has a little bit of complicated math with the bottler and with the company. He said, "You need to know the bottler's margins versus the company's. You need to know the brand margins." It was hard. Every day I went into the job, I felt like I was sweating bullets, but I was learning.

The reason why I want to mention this and go deep is because Gary said to me one day, "Tell me what you think matters in the business." I started talking about consumer purchase power and whether they like the taste of our brands. He looked at me, and he said, "Wow. I thought I trained you better than this." He said, "You've missed the point." I said, "What?" He said, "We're going to have great brands. We're going to have great products because we have a lot of great people making them."

He said, "But if a community doesn't want you in their community, and you use up the resources that are there for them, they will not welcome you in. Don't ever forget that the community is the reason why we're successful. We will continue to have great brands, but if there's no one there who wants you in their community, it won't matter how great they think your products taste."

He said, "We're a lot bigger business than you realize in terms of the emotional connection. If you were simply a brand to refresh people and hydrate them, we would be just a really great drink. But we're not. People expect more from Coca-Cola, and it's because of what we do in other places."

Fast-forward, and my career went on this amazing journey. I ended up getting promoted, and I ran sports marketing. I had a chance in 2008 to work in China right after the Chengdu earthquakes. It was also an Olympic year. That was the first time I met Irial Finan, the gentleman who was running our Bottler Investment Group. He's phenomenal. He's since retired. He said, "We're going to take you out there and have you work with this guy, Martin Jansen, who's in operations. You're going to help us rebuild this whole area of Chengdu. You're going to have to see what's happened with the crisis, and you're going to probably be sleeping in a tent." I replied, "This is the same way I felt when Gary offered to demote me."

For me, it was another experience where I wasn't out there to sell Coca-Cola products. I was out there to rebuild the community with the people and to make sure they had access to water and supplies. I realized that I had been falling in love with this new world that I'm in today. I thought, "This is amazing." I bring it up because they are moments in my career where I felt I started to really learn what Coca-Cola was and that helping people was in our company's DNA.

What I realized over my career, particularly while leading the sustainability strategy, is that I was answering two core questions. How will you make a meaningful difference? How do you make sure that what you're doing drives the scale of change that the community needs, which then ultimately benefits the business as well? I was always drawn to sort of tying these things in, but now I was working at a company that really believed in its core DNA that we could make a difference.

Pagitsas: Bringing us to the present day of your career at Coca-Cola, how do you partner with James Quincey, chairman and CEO?

Perez: It was interesting with James because he took us even further. I remember working with James when he was the head of Europe and even before that as the head of Great Britain. So, I've worked with him off and on, on different sustainability initiatives over the years. He always gave me very good guidance even then.

When he came in as CEO in 2017, he asked me to expand my role and take on more. He said, "I am asking you to take on more so you can further integrate sustainability into the business." At the same time, he also gave me all these sponsorships. He knew I had done sports marketing in the past. But he said, "I want you to truly embed sustainability into the next Olympics that we do, the next World Cup, the next whatever it is. I want this to be embedded in. I don't want to see just a commercial program."

It was one of those moments where I stepped back. I thought, "Gosh, he's right." James is the one who took it deep into the operator compensation. He's the one who pushed that part forward and made sure that it was everywhere you look and touch, from how we design new brands to when we do mergers and acquisitions. It's one of the key criteria we look at for a merger or an acquisition. If they need to improve, there'll be time to mitigate.

He's the one who really took it to that next level. I call it the difference between standing on an Olympics stage as a silver medalist versus a gold medalist. James is the one who said, "I expect top Olympic-gold performance in this sustainability space." That's what he's done. It's been remarkable because after being in the role for a while, I thought we'd been doing pretty well. James was the one who helped me figure out where we could do better. And while we have a lot to be proud of, we recognize the journey continues!

Ezgi Barcenas
Chief Sustainability Officer
Anheuser-Busch InBev

Ezgi Barcenas is the chief sustainability officer for Anheuser-Busch InBev (AB InBev), a publicly traded company based in Leuven, Belgium, with secondary listings on the Mexican Stock Exchange and the Johannesburg Stock Exchange and American depositary receipts on the New York Stock Exchange. A multinational drink and brewing company operating in nearly 50 countries and with over $46 billion in revenue in 2020, AB InBev has a diverse portfolio of more than 500 brands. Since joining the company in 2013, Ezgi has held key roles within the corporate affairs and procurement functions, including global vice president of sustainability. She was promoted to chief sustainability officer, a newly created C-suite role, in August 2021.

In her role, Ezgi coordinates across all business functions to advance sustainability and inclusive growth. Her team focuses on agricultural development, water stewardship, circular packaging, climate action, and responsible sourcing to build a sustainable, inclusive, and resilient value chain. She leads ESG engagement and reporting and oversees the 100+ Accelerator, an award-winning program that pilots and scales sustainable innovations.

Before joining AB InBev, she worked in foreign trade, public health, and international development. In these roles, she witnessed firsthand the role of business in the economy and society and the materiality of sustainable development for business and value chain resilience.

© Chrissa Pagitsas 2022
C. Pagitsas, *Chief Sustainability Officers At Work*,
https://doi.org/10.1007/978-1-4842-7866-6_2

Ezgi serves on the AB InBev Foundation board, the Adweek Sustainability Council, and the Circulars Accelerator advisory group.

Chrissa Pagitsas: Ezgi, tell me about sustainability at AB InBev.

Ezgi Barcenas: As the world's leading brewer, we have operations in nearly fifty countries and sell our products in over 150 markets around the world. We like to think of ourselves as a "global-local" company. Most of our products are sourced, brewed, and enjoyed locally. That gives us an unparalleled connection to local communities, supply chains, and economies, which sit at the core of our sustainability strategy.

At AB InBev, we are driven by our purpose to "dream big to create a future with more cheers." To us, a future with more cheers is shared prosperity for our communities, our planet, and our company. From building a resilient and agile value chain to solidifying our role as a trusted partner in local communities to identifying and capturing new sources of business value, sustainability plays a key role in delivering our company strategy and purpose.

Furthermore, our corporate sustainability efforts allow us to respond more authentically to the rise in conscious consumerism. Especially coming out of the pandemic, we're seeing more consumers becoming environmentally conscious and socially aware. Similarly, our colleagues are increasingly looking for that shared sense of purpose and pride in the company they work for. Sustainability is a genuine concern for the modern workforce and is not only critical for talent attraction but also for talent engagement and retention.

Last but not least, sustainability for us is about innovation. It allows us to rethink and redesign the future of packaging, the future of agriculture, the future of logistics, etc. We see sustainability as the "ultimate design brief."

Pagitsas: Where was AB InBev's sustainability strategy when you first joined, and what is its scope today?

Barcenas: When I first joined the company in 2013, I was working on sustainability within the corporate affairs function. The role was a great springboard for me to quickly learn about our company, industry, and the future of our business. I was also fortunate to have had some of the best managers and colleagues who taught me their leadership philosophy and values.

At the time, we mostly focused on sustainability activities within our own operations, such as water efficiency, energy efficiency, and recycling and waste management. About four years later, in 2017, we moved the sustainability function to the procurement organization reporting to our chief procurement and sustainability officer.

That new organizational design gave us a lot of ability to integrate sustainability into the supply chain. We started having more strategic conversations with our suppliers around how do we rethink the future. What is coming out of our innovation pipelines? How do we co-innovate to tackle shared challenges? It changed the dynamic from a more transactional relationship to more collaborative engagement. It gave us that ability to truly operationalize sustainability across procurement, logistics, and supply teams, which is our entire operations team.

To scale and accelerate impact, in 2018, we launched Eclipse, a collaboration platform for suppliers designed to advance sustainable practices at pace and scale. Eclipse brings AB InBev's vast network of suppliers and partners together to collaborate and make progress on shared sustainability goals. The platform offers educational webinars and workshops with sector experts and provides opportunities to share best practices. Recent supplier collaborations have led to the launch of low-carbon cans, the introduction of alternative fuel delivery trucks, and eco-friendly coolers, for example. This strategic collaboration with our supply chain, in turn, earned us a spot on the CDP 2020 Supplier Engagement Leaderboard, ranking among the top seven percent assessed for supply chain engagement on climate change.

Pagitsas: At the time of this interview, you were just named AB InBev's first full-time CSO. Congratulations. Tell me how this role came about.

Barcenas: The ESG [environmental, social, governance] agenda emerged from 2020 as more urgent than ever. The COVID-19 pandemic accentuated the materiality of ESG performance and its importance to business continuity and value chain resilience. In August 2021, I stepped into the newly created chief sustainability officer role—a fully dedicated C-suite role reporting to our recently appointed CEO, Michel Doukeris. Elevation of the sustainability agenda reflects our commitment to driving business outcomes and delivering shared value through our ESG efforts.

Specifically, the creation of this role allows us to take what we've done on the operational side of the business to create resilience and brings it closer to the corporate strategy to build a more nuanced, forward-looking ESG agenda that will enable our commercial vision. It's been an incredible journey. I'm thankful for the commitment and drive of our team of teams around the world and for the support of our CEO and our board.

Pagitsas: On this journey for AB InBev, it seems that there were two critical points where sustainability was brought closer to the business. First was the movement of the sustainability team to the procurement division, and then there was your appointment to the C-suite and reporting to the CEO.

Given your experience in three different parts of the AB InBev organizational structure—corporate affairs, procurement, and executive management—does it matter where the sustainability leader sits within the organization? How should the sustainability function partner with the executive team, the heads of the businesses, and operational functions?

Barcenas: Great question. When you're sitting in a sustainability role, it is less important where the reporting line is and more important whether you have the buy-in from your CEO and the board. In our case, we've always had that. We've been very fortunate to have had a board and a CEO who understand and embrace the value of sustainability and the materiality of ESG. Carlos Brito, our former CEO, was a big supporter of our sustainability agenda, as is our current CEO, Michel Doukeris.

With such support, my role allows me to be in the room when we're discussing growth strategy. Now I can play a more deliberate role in building a sustainability agenda that enables the commercial vision. It gives me a fundamental opportunity to promote new ideas, discuss stakeholder expectations, build internal consensus, and help adapt to a rapidly changing external environment.

I firmly believe that if you're heading up sustainability, you need to have a belief system that a better world is possible and a worldview that is long-term and systemic so that you can really bring in that big picture. No matter where you're sitting within the organization, your peers and other leaders are always going to prioritize their own functions or departments. As a sustainability leader, you must look across the whole enterprise, create spaces for all those leaders to come together, and build a shared understanding of the problem and the opportunity. That's the real mandate of the role, regardless of where you sit in the organization.

Pagitsas: I couldn't agree with you more. Regardless of where the CSO sits, the executive management, business, operations, and sustainability teams have to have a shared commitment that sustainability will be a part of the enterprise's core strategy and integrated into daily business decisions.

How do you operationalize that integrated approach at AB InBev?

Barcenas: Yes, absolutely. We always had a core sustainability team. However, we also built a "team of teams" around the world and across different functions that are responsible for the execution of our strategy. As a target-driven organization, we treat sustainability like any other business initiative. We set targets tied to our variable compensation for our teams to create a shared sense of accountability and alignment on the mission. That "team of teams" has always played a big role in how we approach sustainability. Now we are working to build a first-of-its-kind community-driven function.

This is not about sustainability having an entirely separate function or organization within the business. Again, this is about how you work across those disciplinary boundaries and beyond the divides so that you're able to further integrate sustainability across the operating model for long-term success.

Pagitsas: Will your approach to the "team of teams" change as you continue to embed sustainability into AB InBev? What key questions will you be asking of yourself, your team, and the company?

Barcenas: My team will remain a team of teams. I will continue to work to empower and enable teams across the company to make an impact in their everyday job on sustainability, whether they're an energy and fluids director, an agronomist, a packaging specialist, a brand manager, a corporate affairs lead, or an innovation VP.

The key question I will be asking my team is, "How do we continue to raise the bar for what it means to be a next-generation business today?" I think that's the next big wave in sustainability—how companies will truly transform the way they craft strategy, reinvent their operating models for long-term success, and reimagine the value proposition for all their stakeholders.

Pagitsas: Are consumers in different countries or geographic regions expecting different actions from AB InBev when it comes to sustainability?

Barcenas: Yes. We definitely see that difference when we do consumer insights work. Depending on the market, depending on the regulatory environment, the consumer behavior, and the market maturity, you see that variation. As part of our refreshed ESG strategy, we have sharpened our focus on eight strategic ESG priorities—water stewardship, sustainable agriculture, entrepreneurship, climate action, circular packaging, diversity and inclusion, smart drinking and moderation, and ethics and transparency.

And we have identified three cross-cutting themes—inclusive, natural, and local—that encapsulate who we are and how we create shared value. Across all priorities, these signature themes are what distinguish us, and their power becomes clear when applying them in different markets.

In developing markets such as Brazil and Mexico, we intend to grow inclusively by empowering small businesses with access to technology, financing, and skills. In developed markets such as the United States and Europe, it is by investing in nature and reconnecting consumers to simple ingredients that will "premiumize" our portfolio. And in emerging markets like Uganda, Nigeria, and Tanzania, it is by strengthening local economies and supporting smallholder farmers that we will build our category.

These allow us to be authentic and create programs and messaging around what is most relevant within the local context and what better targets our consumer needs. We can't lose sight of the local context and the work we still have in front of us.

Pagitsas: How did you and your sustainability teams set the goals that teams around the world are implementing?

Barcenas: Our 2025 sustainability goals drive our short-term actions. Our ambition to achieve net zero across our value chain by 2040 allows us to pursue leading-edge innovation that will become technologically viable, scalable, and economically competitive over time. To design our 2025 sustainability goals, in 2017, I brought together about seventy colleagues from around the world representing procurement, supply, logistics, research and development, and corporate affairs. You'll notice they are all on the operational side of the business. If I were to do it again, I would also invite our commercial colleagues so they could voice changing consumer needs, catalyze product innovation, and help unlock more authentic storytelling. That's the real opportunity we have now as we further elevate and evolve our sustainability agenda.

Pagitsas: What was the outcome of these conversations?

Barcenas: We started looking deep into our business and our value chains to ask, "What are the big opportunities for us to innovate? What are the ways that we can create resilience or formalize those local supply chains?" Out of that, we came up with our 2025 sustainability goals that are organized across four pillars—water stewardship, smart agriculture, circular packaging, and climate action. We took an "outside-in approach" so that we could best design the initiatives that are going to transform how we think about our business and our value chain and how we engage with our partners. We also invited external partners to provide feedback and strategic guidance during the goal-setting period.

For example, our water commitment is that, by 2025, 100 percent of our communities in high-stress areas will have measurably improved water availability and quality. The key part of this commitment is the "measurable improvement" piece. We've made huge strides, especially in the food and beverage industry, around water stewardship. However, when we were designing the goal, we looked at the watersheds that we've invested in over the years. We then asked ourselves and our NGO [non-governmental organization] partners on the ground, "Can we say the health of this watershed is better off now?" The science wasn't quite there. We have to go deep into the local context-based metrics and understand the local indicators of improvement. Water risk is very local. Water security is very complex. We wanted to create shared accountability around measurable improvement with our partners. This approach was welcomed by many water experts.

Next with our agriculture goal, our commitment is to have 100 percent of our direct farmers be skilled, connected, and financially empowered. It's our brand promise to the farmers, an aging population around the world. We source barley, sorghum, cassava, corn, rice, and hops directly from over 20,000 farmers in thirteen countries. Whether they are smallholder farmers in Africa or commercial farmers in Argentina, we provide access to crop varieties, training, and timely insights for better decision-making and financial tools to build resilience. We're also investing in agronomic guidance and remote field-level advice to improve soil health and farming practices, lowering our environmental impact while improving farmer productivity. We're connecting them to other farmers through technology to benchmark crop yields, weather data, and market data. Through BanQu, a blockchain-enabled supply chain platform, we're providing more than 5,000 smallholder farmers across Tanzania, Uganda, Zambia, Colombia, and Ecuador improved security of payments and a digital economic identity to access formal financial services while increasing traceability in our supply chain. We wanted to create a farmer-centric commitment because we know that if we can solve the farmers' problems, we can also solve our supply security challenges. That, for us, was a big insight.

As champions of circular economy, we aim to have 100 percent of our products be in packaging that is returnable or made from a majority of recycled content by 2025. By the end of 2020, nearly seventy-five percent of our products were in packaging that was returnable such as kegs and returnable glass bottles or made from majority recycled content such as the cans portfolio with more than fifty percent recycled content.

The returnable, two-way bottle system follows the old milkman model. In markets like the United States, it's disappeared almost entirely, except for small pockets of the country. In many markets around the world, we have a unique ability to protect and promote that volume. It's a lower price point for the consumer, a higher margin for us as a business, and better for the environment. This is because you can use that bottle up to thirty to forty times, depending on the market conditions. The two-way bottle system reduces your virgin material use, eliminates waste, and lowers your carbon footprint. Similarly, we continue to leverage internal know-how and supplier expertise to test and scale technologies to increase recycled content in our package. This approach helps us further eliminate packaging waste, reduce the need for virgin materials, and decrease carbon emissions.

Our last commitment is to climate action. In addition to our newly launched ambition to achieve net zero across our value chain by 2040, we have short-term climate goals. We're proud of our commitment to source 100 percent of our purchased electricity with renewables by 2025. We aren't looking to offset our electricity usage at our plants by purchasing renewable energy and carbon offset credits. Instead, we are trying to increase renewable energy capacity around the world through additionality. In markets where we have the size and scale, we have an opportunity to help transform local electricity grids with our demand and learnings.

We were one of the first 100 companies to set a science-based target. Against a 2017 baseline, by 2025, we're looking to reduce greenhouse gas emissions per beverage by twenty-five percent across the whole value chain, inclusive of scopes 1, 2, and 3, and reduce absolute emissions in our operations by thirty-five percent for scopes 1 and 2. We are on track to deliver these commitments and continue to pursue and scale innovation for deep decarbonization.

Pagitsas: I appreciate the focus on quantification. It is important for leading companies to do that if they want to be seen as credible. Once you identified the commitment areas and the targets, how did you get buy-in from senior leadership on the targets?

Barcenas: Throughout the goal-setting process, it was important to bring everyone along with us on the journey and create that buy-in early. This naturally happened, given we had invited a cross-functional set of colleagues to join the goal-setting process. They were an integral part of the strategy development. At the same time, we created a sustainability council made up of four chiefs. We would update them and our CEO regularly on the strategy and goal-setting process. Now we have six C-suite members across different functions that oversee sustainability governance.

Right before launching the commitments publicly, I remember in one of our leadership meetings, we were asked, "All these targets sound great. They make business sense. But how confident are you that you're going to get to 100 percent by 2025?"

Pagitsas: How did you respond, Ezgi?

Barcenas: We talked about the cost of leadership and how sometimes we may have to set inspirational or moonshot targets in order to find new pathways and set a new vision. But I was very honest and transparent. I said, "We know how to get there eighty-five to ninety percent of the way, but for the remaining ten to fifteen percent, we're going to have to find new ways of working, new partnerships, and new innovations to tackle these challenges. However, we don't want to set an eighty-five or ninety percent target. We want to set a 100 percent target and work to get there even in difficult markets and challenging environments." The follow-up question from leadership was, "Okay, great, how are we going to do that?"

We had a few cross-functional leaders in the room who were very supportive of the agenda and unanimously decided to create a sustainability accelerator there and then. That decision has not only been critical to meeting our goals, it has also paved the way for what was to become one of our signature programs, the 100+ Accelerator.

We launched our goals in March 2018. By August, we launched our 100+ Accelerator program, which has been incredibly successful. We've accelerated thirty-six startups in sixteen countries in the first two years of the program.

As of 2021, we're hosting the third cohort with new corporate partners. We have joined forces with Unilever, Coca-Cola, and Colgate-Palmolive to identify and pilot new innovations with thirty-five startups across over twenty countries. No one company can solve today's sustainability challenges alone. Through these kinds of collaborative initiatives, we can work together to catalyze sustainable innovation while tapping into the world's brightest minds. There's a ton of innovation out there.

Pagitsas: Would it be safe to say that there had to be a leap of faith in setting these commitments? All the technologies that you and the accelerator companies need to achieve your commitments didn't exist when you set them, but by believing in the power of innovation, you will create them.

Barcenas: Yes, that is correct. The other big factor for us was having the comfort of knowing that we knew and understood our business and our value chain. Without that, you can't design an impactful sustainability program. You need to understand the ins and outs of the business. You need to make sure that the management and technical teams are helping you design what an authentic and impactful sustainability program should look like.

Pagitsas: Let's talk about the COVID-19 pandemic. What impact has it had on AB InBev's business?

Barcenas: We first experienced the pandemic in our China operations. Our teams in China demonstrated great agility in responding to the pandemic and shared lessons learned. As the pandemic moved over to Europe, Africa, and eventually the Americas, we kept learning from each other.

One of the biggest challenges we tackled with our stakeholders was how to ensure we have supply security but at the same time do it safely. For example, how do we keep barley buying centers open to ensure that cash flow to our farmers? It emphasized the need to find new ways to engage with suppliers and our customers, the retailers, alike. In 2020, we produced and donated more than four million bottles of hand sanitizer and disinfectant to hospitals and frontline health workers in more than thirty countries. We provided water and medical supplies to frontline emergency workers around the world, including the donation of three million face shields in Brazil. We mobilized our truck fleets in markets such as Colombia, Peru, and Ecuador to deliver essential food and medical supplies to areas where they were most needed in coordination with the government agencies in charge of responding to the COVID-19 crisis. We donated billboard space intended for our brand campaigns to be used for important public health messages in Belgium, France, and the Netherlands. The pandemic has shown us that we could serve in new ways to make a positive impact as long-standing partners of the communities we live and work in.

Pagitsas: How did COVID-19 shift, if at all, AB InBev's and other company's view of sustainability and ESG?

Barcenas: Around March and April of 2020, I received many questions from external stakeholders. Does the pandemic mean that the sustainability strategy will be put on the back burner? What does it mean for the UN Sustainable Development Goals? Are we going to fall behind on our commitments? My sustainability peers at other companies were getting similar questions from their stakeholders.

Fast-forward to November and December of 2020, and there was this huge recognition across the board about the importance of ESG and what it means for business continuity, value chain resilience, and community wellbeing. This was a defining moment for the rise of ESG within the capital markets. That was a notable shift for sustainability agendas worldwide.

Pagitsas: What is an example of attention from the capital markets?

Barcenas: We have seen a surge in sustainable finance. In early 2021, with the recent rapid maturity of the sustainable finance market, particularly the SLL [sustainability-linked loan] structure, combined with our need to refinance our revolving credit facility [RCF] because of the upcoming maturity, we announced a sustainability-linked, $10.1 billion RCF, reinforcing our commitment to sustainability. As the capital providers continue to set their own ESG goals, we will likely see more momentum in this space.

Pagitsas: What do you anticipate will change in business as a result of COVID-19?

Barcenas: Another big learning from the pandemic was that the lines around environmental, social, and economic sustainability have become blurrier. For the first time, we could see how a biodiversity loss could create a huge impact on public health in the form of a global pandemic and how livelihoods would be impacted through that. All these topics are intimately interconnected.

It is my deepest hope that this will continue to be recognized so that we no longer try to tackle these challenges in silos. You cannot just solve a biodiversity or an ecosystem problem without thinking about the water access challenges or thinking about the livelihoods of the farming communities. That learning has inspired a big paradigm shift, and sustainability strategies need to embrace the fact that we have to "multi-solve" if we want to create a sustainable, inclusive and prosperous future.

Pagitsas: What drives you as a leader to tackle these difficult challenges at the intersection of sustainability and business?

Barcenas: For me, it's always been about making progress both personally and in terms of the broader society. I think my upbringing influenced this mindset. I was born and raised in Cyprus. My parents were instrumental in encouraging me to always think about progress and how I could aim for a better education and a greater purpose.

That led me to study engineering in the United States. To this day, I still lean in on my curiosity and design intuition, I love a good challenge, and I'm always up for learning something new. It was important to me coming out of school to find a field where I could see the immediate impact of my work and drive that progress. I feel privileged to have found my passion early in my career, and I am fortunate that I get to do this every day with a diverse and dynamic team that shares the same deep sense of purpose, urgency, and optimism to create a future with more cheers.

Pia Heidenmark Cook

Chief Sustainability Officer
IKEA

Pia Heidenmark Cook was the chief sustainability officer at the Ingka Group, the largest IKEA retailer through July 2021. She led IKEA's strategy to become people- and planet-positive by 2030 and was inspired by the UN Sustainable Development Goals (SDGs) to create a better everyday life for many people. IKEA is a global home furnishing brand with 456 stores in 61 countries and approximately 40 billion euros in revenue in 2020.

Prior to her appointment to the role in 2017, she served as head of sustainability for the retail unit, and she joined IKEA as head of communications for the IKEA Foundation in 2008. Pia has extensive experience internationally in the retail and hospitality sectors. Her deep experience in sustainability was grounded in serving as vice president of Responsible Business for the Rezidor Hotel Group and as chairperson of the International Tourism Partnership.

Pia serves on the board of trustees for the Sustainable Hospitality Alliance and the boards of directors for Scandinavian hamburger chain MAX Burgers AB and US-based sustainable materials company Origin Materials. She has previously served

© Chrissa Pagitsas 2022
C. Pagitsas, *Chief Sustainability Officers At Work*,
https://doi.org/10.1007/978-1-4842-7866-6_3

as co-chair on the Retailers Environmental Action Programme with a member of the EU Commission. She has been recognized as a top five influencer on climate change by the Climate Group, a top 100 climate influencer on Twitter in 2020, and a top ten female leader in sustainability by World Business Council for Sustainable Development in 2018.

Chrissa Pagitsas: Let's dive into the IKEA sustainability strategy. What is its scope?

Pia Heidenmark Cook: The IKEA People & Planet Positive strategy describes the sustainability agenda for the total IKEA value chain. Our sustainability ambitions and commitments are set for 2030 in line with the UN Sustainable Development Goals, and we want to have a positive impact on people, society, and the planet.

We have identified three major challenges that are highly relevant for our business—climate change, unsustainable consumption, and income inequality. In response to these major challenges impacting the IKEA business, we have identified three corresponding strategies to help us fulfill the IKEA vision and ambition to become people- and planet-positive.

Let's first focus on our strategy to support healthy and sustainable living. Our ambition is to inspire and enable more than one billion people to live a better everyday life within the planet's boundaries by 2030. Our homes and the way we live have a huge impact on our health, well-being, and the planet. We want to help make a more sustainable life both affordable and attractive.

A tangible example of this strategy is our development of home furnishing products that enable our customers to live more sustainably at home. We promote these products through all sales channels, from store displays and room sets to IKEA.com. For example, IKEA now sells residential solar panels in fourteen countries, water-efficient water faucets, smart waste sorting bins, and energy-efficient LED [light-emitting diode] lights. These products enable our customers to reduce their own environmental footprint at home. In addition, IKEA works to minimize the footprint of its products through using more sustainable materials, such as renewable and recycled materials. Ninety-eight percent of the wood used in IKEA products is either certified by the Forest Stewardship Council or recycled.

Next is our strategy on supporting a circular economy and being climate positive. By 2030, our ambition is to become climate positive and regenerate resources while growing the IKEA business. We have only one planet with limited resources. Pressure on forests, fisheries, and agriculture, loss of biodiversity and wildlife, ocean pollution, erosion of soil, and increasing levels of air and freshwater pollution affect the lives and livelihoods of millions of people around the world. An example of how we are working to reduce our greenhouse gas emissions is by reducing energy use and switching to renewable energy across the value chain. In the last two years, we have grown our overall business while reducing, in absolute terms, our greenhouse gas emissions. Thus, we have started to decouple growth from climate change.

Being fair and equal is our third strategic focus area. By 2030, our ambition is to create a positive social impact for everyone across the IKEA value chain. We believe that through our business, we can support and influence positive change. With our size and scope comes both responsibility and great opportunities to create positive change for many people. Our longest-standing work is connected to IWAY, the IKEA Code of Conduct for suppliers. It was launched back in 2000 and sets environmental and social requirements for our suppliers and requires IKEA purchasing organization to work strategically with suppliers.

Pagitsas: Let's dive deeper into one of the three pillars of the IKEA sustainability strategy—healthy and sustainable living, circular and climate positive, and fair and equal. Which do you want to address?

Heidenmark Cook: Healthy and sustainable living because the core of our business is life at home. That's what we do. Conceptually, healthy and sustainable living is fundamentally about how we shift home furnishing products, life-at-home solutions, and know-how to fit within the planet's boundaries. It means thinking about the whole product development cycle and the supply chain. We have to ask ourselves, "What goes into the product? How do we design the products? How do we source the materials from our supply chain?"

The products are one big part of the equation, but the other part is the customer-facing strategy both in the store and on the web. We must ask ourselves, "How do we tell the story to our customers that these products have been made from recycled material, from renewable material, or maybe from an innovative material? Or that we've removed certain substances and become more stringent on the content?" We may also want to communicate to customers that a product has increased functional capability. For example, if a customer buys an IKEA LED light, they won't waste a lot of energy and can just enjoy the light. If they buy an IKEA water faucet, it uses fifty percent less water than another faucet in the market, and you save both water and money.

It's that story of both functionality and sustainability that we want to tell our customers. It's one of the most complex and complicated things to do because people come to the store because they have a need at home. They may walk into our store thinking, "I need a chair, I need a sofa. Or I need inspiration because I don't know what to do with my living room." They are not necessarily thinking about product sustainability first.

However, we want to tell the story that everything you buy from IKEA goes through our stringent process of IWAY. It doesn't matter which supplier provided the product's materials. We put the same code of conduct in place.

We don't have a separate designation for all the "creative, innovative material products" and then a designation for "the rest." We work across all suppliers providing all the metal, all the cotton, and all the wood for our products.

But it is a challenge to tell this story. I used to sit on several IKEA retail country boards, which performed country-specific research on sustainability. Their research concluded that if you, the manufacturer or service provider, tell too little, then consumers may assume you're hiding something, and you're not transparent. If you tell too much, then consumers assume you're greenwashing. So how do you find that golden middle ground where you're credible? It's so difficult. I don't think we have found the solution yet, but we are way better than we were some years ago.

At IKEA, we are trying to shift our communications strategy so that the sustainability of our products really stands out. We're working to increase our in-store communications with banners and other signage and provide more training for coworkers. If you are a customer and were to say to one of our store coworkers, "I want to buy a new bed," the coworker on the shop floor should be able to tell you our mattresses are made of more sustainable cotton—often from the Better Cotton Initiative, they are good for your health, there are no bad chemicals in our mattresses, and at the end of its life, we'll take that mattress back, and it will go to recycling.

I do think we are telling a better story today than we did a few years ago. And I'm sure we will tell an even better, more sophisticated story in a few more years. Finding how you communicate the sustainability story in a way that resonates with people—but without overdoing it and overclaiming it—is a work in progress for us and many other companies.

Pagitsas: IKEA is a global company with stores in Asia, Europe, and the United States. Is the consumer asking for something different from IKEA's sustainability story and strategy in the United States versus Europe, Russia, or China, for example?

Heidenmark Cook: Definitely. The conversations around the water cooler in the United States are not as much about sustainability as they are around the coffee machine in Europe. It's changing in the United States, but awareness of sustainability in Northern Europe is quite high. In Europe, the conversation is about where the product comes from and who made it. There seems to be a greater understanding of the sweat and tears that went into the manufacturing of this product. The European thinking is more along the lines of "I need to take care of this product, and I just can't waste it."

In Russia, the sustainability conversation focuses more on the "here and now" environmental problems. So, the visible and immediate problems of waste and water contamination. In China, the conversation is more about how the products impact health. So, I would say consumers are focused on different aspects of sustainability depending on the country or region they live in.

Pagitsas: How do you tailor your products to meet these diverse consumer demands?

Heidenmark Cook: We create a global range of home furnishings products. Today, we sell over ten thousand products at our stores and on our website. We source more products for different geographic regions, and we localize the range partly. We adjust for local market preferences. For example, we do a lot of local marketing research to see which products we should lift in the market. Or which sustainability story resonates the most with consumers. But, if you go to the IKEA store in Philadelphia and one in Shanghai, they more or less look the same and have the same products.

As a result of the different geographic and market preferences, I can't say that I have the same sustainability expectations for every country with regards to local actions. However, they all need to add up to the global sustainability goals. One area that we are setting as an expectation for all our stores globally is to buy renewable energy no matter if they are in a country where it's more difficult or more expensive. We know that the journey to get there requires more time and support in some countries.

In some countries, it's easy. In Sweden, we can buy electricity from the electricity grid, and it's green. In other countries, we take different approaches. For example, in Russia, we invested in a solar park that will feed all our seventeen stores and shopping centers. We're the first non-energy company and the first international company to make such an investment in Russia. It requires money, stamina, and courage to test a new approach. The implementation approach is very different from country to country, but we still have one sustainability strategy.

Pagitsas: Thinking about the sustainability strategy you built, did the company and leadership adopt it immediately? Or did you hit barriers?

Heidenmark Cook: That's a very good question. I feel like I get *no* all the time. I would say the first time you come to your peers or leadership with an idea, there's a lot of *no*. In the first iteration of our sustainability strategy, in 2012, we discussed using electric vehicles for home deliveries and new locations for our stores. Using electric vehicles was a *no* in 2012. It was just not possible. We were not ready as a company—or a society.

However, today our strategy has shifted. I've signed up the company to have 100 percent electric vehicles or other zero-emission solutions for all our home deliveries. The *no* became a *yes*. You need the perseverance of knowing what will happen, what needs to happen, and then be patient until the right time appears to move on the new idea.

Pagitsas: Did you need the same type of patience at other companies at which you worked?

Heidenmark Cook: Before IKEA, I worked at Rezidor, a hotel company based in Brussels. My CEO at Rezidor said to me, "I didn't always agree with you, but I knew you knew what needed to be done. Then you stuck to it." When I left after seven years, I looked at my checklist. I wrote it in the first half of the year that I joined Rezidor. I'd ticked off some of the items on the checklist, but some were not ready to be tackled while I was there. But they've done them now. It is all about change management. You need to bring people with you, and they need to be ready.

Pagitsas: Looking at the role of the chief sustainability officer (CSO) at large global companies, few include running a line of business and leading the sustainability strategy. Why aren't those scopes joined together more frequently within companies?

Heidenmark Cook: I think they can be, and increasingly they will be. I'm attending an INSEAD business school program for executives serving on company boards called The International Directors Programme. We had a value creation session focused mainly on finance. The INSEAD professor talked about the perceived opposition between profit and purpose. Business leaders, as well as the larger society, still assume that working with sustainability is not profitable and that investors are not interested in sustainability.

This professor's theory is that this assumption is no longer valid. However, we still sit with a feeling that that's how it is. So, there is something in that notion of thinking that profitability and purpose are opposing, and that's why you can't be both the head of a business unit and CSO. One does the "real stuff," the business, and puts bread on the table. The other one does the "good stuff."

I think that the two roles will merge, but it's going to take a while because today, there is still a false narrative around the scope of the two roles. On the business side, some people may think, "Here's the side of the business where you think short-term, you make the tough decisions." With the sustainability side, the same people may think, "Here's the side of the business where you're thinking long-term, about people." Therefore, they justify the separation of the roles because of the short-term versus long-term timelines also known as the tragedy of the horizon. But I don't think they should be separate.

Pagitsas: It is true that non-sustainability business leaders generally have quarterly and annual targets to deliver on. In contrast, sustainability leaders such as yourself think about targets into the next decade or two. In 2018, IKEA set goals to be achieved by 2030. How do you and your peers who lead the business side at IKEA manage the different timelines?

Heidenmark Cook: At IKEA, we're retailers. We run on weekly timelines, daily, even hourly. The focus is always about knowing if the shelves are full, so products can go in and out. For example, an SMS goes out to staff globally about the latest week's sales result. On the shop floor, there are daily or hourly follow-ups on sales results. We work on annual financial budget years. Traditionally, our strategic business plans have been yearly, but we have started working more on three-year plans.

When I first introduced the sustainability strategy, which reached to 2030, everyone was saying, "Whoa, we're stretching ourselves!" In response, I said, "We have three more three-year plans, and then we are at 2030. So, what needs to happen now to enable the next three-year plan and then the next?" By approaching it this way, the 2030 goals became short-term deadlines. That has helped the organization to say, "2030 is not that far off."

It is demanding to manage the business "here and now." When you talk to a lot of people in an "open heart conversation," they care, and they want to think about sustainability. But then they're caught up in the day-to-day events and are in meetings back-to-back, addressing people issues, managing sales dropping, and solving a crisis. However, I understand these pressures and know that sometimes we cannot meet immediately to develop a sustainability solution.

Pagitsas: In addition to the short-term versus long-term timeframe difference, there seems to be another dimension that is different between traditional business leaders, which is the unit of measure of success. As the CSO, you're driving IKEA "to inspire and enable more than one billion people to live better lives within the limits of the planet" by 2030. That goal addresses people's experience and the planet's natural resources. In contrast, another colleague may be driving toward a goal of a particular profitability per square foot per store. Are these two goals going to remain siloed, or will they become more integrated into business decision-making?

Heidenmark Cook: I think they are siloed now, yet I think it will be less so going forward when you start working with value creation goals. Up to now, I would say the focus has been on the top line and bottom line. Those are the two KPIs [key performance indicators]. However, when I'm in the room when people are defining the top-line and bottom-line goals, I'm able to say, "Those are not the only goals. We have people goals, and we have customer goals." We're starting to change and look at how we best contribute to customers, coworkers, the community, the environment, and the company's bottom line, but that's the transformation we're in.

For example, we have come quite far with climate and thinking about it as a primary goal. We can now look at the outcome if we put in a dollar here and determine if we are going to get this serious carbon dioxide reduction. We can also analyze the outcome from our investments in renewable energy, for example, and whether we will get the same dollar out when putting it in this country versus that country. This level of analysis has taken a while because you need to take carbon dioxide data together with the finance data at the start of the project to create better, more insightful conversations.

It takes time to build the competence or knowledge to understand that the decisions I make on the sustainability side of the business also impact the company's overall financial goals. We had a session recently on climate with our top leaders. Climate is a business language, just like finance is a language, so it's no surprise that some of the leaders said, "It's so complicated and complex." My response is, "Yes, but everyone can study it." We'll get there, but it is about timing, it's about the KPIs, and it's about learning a new language.

Pagitsas: Are there other areas in which your day-to-day work is different from your senior executive colleagues at IKEA?

Heidenmark Cook: I spend a lot of my time engaging with people and partners outside of the company. Most of my colleagues spend most of their time inside.

Pagitsas: How would you contrast your "outside" job with your colleagues' "inside" job?

Heidenmark Cook: In some companies, as CSO, you report to the marketing and communications team or to corporate affairs as part of the reputation team. I report to the business and the deputy CEO. With this organizational design, a big part of my job is to provide the business and the deputy CEO with insights and foresight on the things that will impact us in the five to ten years that we need to start preparing for today. I need to spend a lot of my time being visible outside. So, I interact with so many other brands and companies and sectors. For example, I sit on panels and speak at conferences with other companies that have sustainability learnings to share. I interact with the partners we work with, such as the World Business Council for Sustainable Development. I have a peer-to-peer dialogue with CSOs in other companies.

I think it's in the nature of the job as a sustainability leader to look for exponential growth and think, "How do we prepare for it?" My entire job is change management. That's what I do.

Pagitsas: Pia, earlier you mentioned the word *outcomes* when talking about carbon dioxide emissions. Are *outcomes* different from *output*? What about the term *impact*? Are they interchangeable?

Heidenmark Cook: No, these terms are not interchangeable. Output is something you can see relatively fast. For example, you launch a good website that is easy to buy from, and you'll have output, which is increased sales. Or you do an activation marketing campaign, and you get people in the store, and you'll have an output, increased sales. Output is kind of relative to what you put in. You can see what comes out, and it's not too far out in the timeframe.

With outcomes, you can ask, "What is the outcome of the sale? Does it create more job opportunities? Does it create something that we can reinvest and do something good with it, like build a new store, support a charity, or invest in renewable energy?" So that's more what *outcome* is about. What does it lead to? Does it have a positive outcome? Then *impact* is even further out and asks, "What is the long-term impact of our actions on the environment and society?" In business generally, we only measure output.

Pagitsas: It sounds like that's where the importance of having foresight comes in, as you mentioned earlier, to create that long-term impact. Let's turn back the clock and talk about your path to IKEA's CSO. Have there been any key turning points in your career?

Heidenmark Cook: There have been a few of those! One of the first was being in business school and then studying environmental economics with a very inspirational teacher. That's where I first learned about unaccounted externalities. This means that there's a consequence of all the consumption, which may not be factored into the price of a good. And consumption impacts the water, the air, the soil, and more. That was an "aha!" moment. After I was done with my business degree, I did a master's in environmental economics, and then I stayed on to do a PhD.

As I worked toward my PhD, I realized that I was tired of studying and analyzing companies and that instead, I wanted to work with companies. I wanted to be part of the "doing" team. As I thought about leaving the PhD program, my professor argued for me to stay since I had one year left until I received my doctorate. However, my response was, "Okay, but who knows if it's going to be a year? I'm ready. I want to go out into the business world and do." In Sweden, you can leave a PhD program and receive a licentiate degree, which is recognized as a pre-doctoral degree. So, I left the program, and that's when I joined Rezidor.

At Rezidor, they had realized early on that sustainability as a business strategy was happening. I built up their sustainability strategy from scratch. The CEO said to me, "I know we need to do something, but I have no clue what we need to do." It was a fantastic seven years building up a sustainability strategy across hotels in fifty countries.

When I wanted to move back home to Sweden, IKEA was *the* company to work for. When I joined IKEA, I realized it was a whole different level of ambition than at Rezidor. At IKEA, we develop and put consumer products into the market, so you must think about the whole value chain. And IKEA had been doing sustainability for much longer. IKEA's first environmental policy was from 1989.

Within Ingka Group, the largest retailer within the IKEA system, our two big business units are the retail division and the shopping center division. Within the retail division across thirty markets, there are approximately 400 customer touchpoints where people can buy IKEA home furnishings products or eat in IKEA restaurants, including stand-alone stores, stores in city centers, pickup points, and IKEA.com. The shopping center division leads our business strategy at large retail developments that house IKEA stores and other brands at approximately forty shopping centers across Europe, Asia, and Russia.

Pagitsas: What's your approach to taking on these large leadership roles?

Heidenmark Cook: I like to have the overview. I like to see how things connect with the management team and understand how pieces fit together. I think also I've always leaned in. If someone asks me to do something, I'm going to say, "Why not? I'll do it." Then I jump in. So, I think that's why I am where I am.

Pagitsas: It sounds like there's an element in your career of not being afraid of the "new." The roles you took on at Rezidor and IKEA and the strategies you carved out were new. There was no roadmap to follow.

Heidenmark Cook: Absolutely. I'm married to an entrepreneur. Compared to him, I'm super square! He builds businesses. Yet, I also like to grow and develop, learn new things, and lean in. I'm not a maintainer of existing businesses and processes. I don't mind painting a picture of what needs to happen. No one needs to paint it for me. I like to paint it and to build it.

Pagitsas: As we conclude our conversation, I'd like to ask you to reflect on leaders in sustainability. Particularly, there seem to be many women who lead sustainability and ESG strategies at for-profit companies. Do you have any observations on why that is?

Heidenmark Cook: There's something about how we lead and why we lead. An amazing woman named Dr. Scilla Elworthy works in peace negotiations. I met her at a conference, and then I reached out to her afterward because she was so cool. She wrote a book called *The Mighty Heart: How to Transform Conflict* [(self-pub., 2020)]. She talks about "feminine intelligence," which is not "female intelligence." Men can have these traits, too. It's fundamentally about courage, deep listening, interdependence, and intuition. She writes that it's something that we often have, although not all women. This feminine intelligence includes understanding both collaboration and interdependence. I think there's something in the nature of working in sustainability that resonates with that skill set.

Pagitsas: Keeping this feminine intelligence concept in mind, how do you lead?

Heidenmark Cook: I never, or very seldomly, work from a position of authority. Sustainability is a cross-functional topic, so you always need to work through influence and inspiration. I need to influence by engaging people and having them want to be part of the sustainability journey.

Even in my own team, which is responsible for the global sustainability function, I need to make the final call as the manager. I choose to lead by trusting my team and being clear on the direction while co-creating the way forward. I simply believe it is a better and nicer way of leading than through power and authority.

Virginie Helias

Chief Sustainability Officer
Procter & Gamble

Virginie Helias is the chief sustainability officer at Procter & Gamble (P&G), a multinational consumer goods company focused on providing branded products of superior quality and value to improve the lives of the world's consumers. The company operates in ten categories where product performance drives brand choice—Fabric Care, Home Care, Baby Care, Feminine Care, Family Care, Hair Care, Skin & Personal Care, Oral Care, Personal Health Care, and Grooming—and had over $76 billion in net sales in its 2021 fiscal year.

Prior to her appointment to this role, Virginie worked on several of P&G's multibillion-dollar brands, including Pantene, Tide, and Pampers. As the Western Europe brand leader for Ariel, one of the company's largest brands, she launched the highly successful "Cool Clean/Turn to 30" campaign, the first P&G consumer-facing sustainability program.

Virginie has extensive international experience, working in France, the United Kingdom, Switzerland, and the United States. She is also a certified coach specializing in change management and leadership development skills.

Chrissa Pagitsas: Procter & Gamble is a globally recognized company with hundreds of brands that every consumer recognizes and has in their home, from Pantene and Pampers to Braun and Tide. What is the scope of P&G's Ambition 2030 strategy as it applies to your business strategy and brands?

© Chrissa Pagitsas 2022
C. Pagitsas, *Chief Sustainability Officers At Work*,
https://doi.org/10.1007/978-1-4842-7866-6_4

Virginie Helias: Ambition 2030 is our roadmap to enable and inspire positive environmental and social impact through our brands across our value chain, employees, and partnerships. It covers all the areas where P&G can leverage its scale, influence, innovation capabilities, and brand leadership to accelerate the development of a circular and regenerative economy. That includes our climate risk efforts, waste program, sustainable sourcing, and regenerative forestry efforts, and water management strategy.

Within climate risk, we have committed to achieving net-zero greenhouse gas [GHG] emissions across our operations and supply chain from raw material to retailers by 2040, as well as to interim 2030 goals to make meaningful progress. We also shared our Climate Transition Action Plan, which outlines a comprehensive approach to accelerating climate action and the key challenges ahead by 2030. At the product category level, our flagship brands, Tide and Ariel, have committed to decarbonizing the laundry process at every step—from the ingredients we put in all the way to how the product is used by the consumer, which is the most important, and finally, to its disposal.

Our waste management program aims to prevent plastic leakage in nature by reducing our use of virgin petroleum plastic by fifty percent, making 100 percent of our packaging recyclable or reusable, and being a catalyst for cross-sector efforts that invest in scalable solutions to improve waste management infrastructure globally. P&G is a founding partner of the Alliance to End Plastic Waste, our CEO chaired it from January 2019 to November 2021.

Our sustainable sourcing and regenerative forestry efforts are focused on supporting the sustainable sourcing of palm oil and wood pulp. We work in partnership with industry and NGOs [non-governmental organizations] to achieve a forest positive supply chain. For every tree we use in our products, at least one is regrown.

Lastly, water is a vital resource for people and for P&G. Seventy percent of our products need water to be used. Therefore, we are working on three different levels to address water challenges at scale. First, we are using the power of water-efficient technology in our products. For example, we have formulated Cascade, the dishwasher detergent, to allow consumers to skip pre-washing dishes. By not pre-washing, you'll end up with cleaner dishes while saving water, as Cascade and your dishwasher work in unison to eliminate stuck-on food and residues. Another example is using our Swiffer WetJet product instead of a traditional mop-and-bucket to clean your floors. This simple could save a household seventy gallons of water per year.

Second, we partner with industry and NGOs with the ambition to protect water for people and nature in eighteen priority water basins globally. We have started a project in the Lower American Basin in California with a goal to conserve or restore 3.3 billion liters of water. Third, we have launched a water cross-value chain effort called the 50 Liter Home Coalition, which aims to reinvent water for urban living. The coalition's goal is to allow people to live well with 50 liters per person per day versus the average 260 liters they use in the United States.

Pagitsas: How are you progressing toward these goals?

Helias: On climate, we have made great progress in our operations—achieving fifty percent absolute reduction of greenhouse gas emissions in the past decade. We have realized this by reducing emissions in our plants and purchasing 100 percent renewable electricity in North America and Europe. That is halfway to our carbon neutrality goal, and we are continuing to reduce emissions as much as possible. However, there is not yet a viable renewable energy alternative for emissions mostly coming from thermal energy. To balance those residual emissions, we invest in natural climate solutions and partner with organizations like Conservation International, World Wildlife Fund [WWF], and the Arbor Day Foundation to protect or restore critical natural ecosystems.

We are in the process of developing a portfolio of projects that not only provides carbon benefits but also creates co-benefits for biodiversity, food security, and the well-being of the communities whose lives depend upon those ecosystems. Identifying those projects requires a new level of collaboration with conservation experts, local governments, and communities to ensure all stakeholders and nature will thrive as a result of these investments.

The bar is rising on the level and scope of cooperation needed to deliver a positive impact at the scale and speed needed to address planetary challenges. System-transformative collaboration is needed. We are still learning how to do that effectively. In many ways, this level of system transformation is unchartered territory.

Let me give you two examples. In the circular economy, P&G established the initiative called HolyGrail—using digital watermarks embedded on the surface of the packaging to enable more accurate packaging sorting at recyclers and, therefore, a greater value of recycled material. This is key to developing the circular economy. This project has progressed to HolyGrail 2.0, driven by Association des Industries de Marque, also known as AIM or the European Brands Association, and powered by the Alliance to End Plastic Waste. It is now moving to a test market phase with a coalition of over 130 players across the entire plastic value chain. We believe it has the potential to revolutionize the recycling industry.

Next, the 50L Home Coalition is also about the decarbonization of homes since lower water levels will make decentralized electricity grids easier. Today's homes and buildings represent one-third of global GHG emissions. The coalition was formed with companies like SUEZ, Kohler, Electrolux, Arcadis,

ENGIE, IKEA, Grundfos, and P&G. It is hosted by the World Business Council for Sustainable Development, the World Economic Forum, and the World Bank's 2030 Water Resources Group to pilot the 50 Liter Home Coalition in multiple cities.

Chrissa Pagitsas: What was your professional journey to becoming the chief sustainability officer (CSO) at P&G?

Virginie Helias: I fell into sustainability a bit like Obelix fell into the magic potion as in an Asterix and Obelix story. It was an accident. Now, I call it *serendipity*, and you will understand why. When I joined P&G in 1988, it was really to work in marketing, and that's what I did for twenty-three years. I had the opportunity to work across multiple multibillion-dollar brands like Pantene, Ariel, Tide, and Pampers and work in different countries, like France, the United Kingdom, the United States, and now Switzerland.

The word *sustainability* was defined in 1987, so it was a very fresh concept when I joined P&G. I didn't even know how to spell it! Maybe it was a coincidence, but the first project I had as a young brand assistant was to launch Ariel without phosphate. It wasn't a sustainability initiative for me. It was just a marketing strategy.

When I later moved to Pampers, I was in charge of launching the compact diapers, which was a major environmental advance. However, to me, it was just flowers and butterflies because I considered myself working on a serious business. It was not until years later, when I was in Europe and leading Ariel for Europe, that I turned to sustainability. Back then, the business was not doing well. I was testing many ideas to revitalize it, and the one that popped to the top was "Ariel cleans so well that you can wash your clothes at a low temperature." This was not a sustainability promise at the time. It was purely about the performance of the product. However, I was working on this challenge in 2006, the same year Al Gore launched his movie, *An Inconvenient Truth*.

After watching the movie, it was the first time that I realized the meaning of global warming. I was struck by the movie, and I took my whole team to watch it. At the same time, I also discovered the scientific concept of life cycle assessment, which is an approach to measuring the impact of the carbon footprint across all the phases of a product, from material extraction all the way to disposal. I learned that for a detergent, the largest footprint impact came from heating the water in the washing machine. That was eighty percent of the detergent's total carbon footprint, from the ingredients to manufacturing to final use by the consumer.

That is when it clicked for me. My "Turn to 30" campaign encouraged consumers to turn their washing machines to thirty degrees Celsius or eighty-six degrees Fahrenheit. It was a major commercial success and, at the same time, it was the best thing I could have done to reduce the footprint of my product. From that day, I saw that sustainability could drive the business and have a positive impact on the planet. I started to volunteer for other sustainability projects. The president of P&G Europe at the time was a big fan of sustainability, so he created an internal sustainability board, and I represented the marketing function.

After a few years of doing sustainability on the side of my day job, I decided to take it to the next level. I attended a WWF seminar called One Planet Leaders at the IMD [Institute for Management Development] in Lausanne, Switzerland. I was surrounded by sustainability professionals. I was the odd one out coming from marketing. By the end of the week, I realized that if I could make sustainability my job, the impact I would have on the business would be so much greater than only being an innovation leader at Ariel. At the seminar, I made a major commitment and said, "I will make sustainability my job," and I had absolutely no idea how I would do that and what that really meant. That's the thing when you have a goal. When you are set, you put things in motion.

Two weeks after the training, I'm flying to Cincinnati and asking to talk to the CEO. I had done a bit of homework, and I had realized that many of the big companies were starting to staff sustainability with business leaders. I basically pitched the CEO for a job that was to be the bridge between the science of sustainability and the business. I said, "I think there is a big opportunity, and, by the way, that's a job I want." He said, "Okay." From there, I started from scratch with the single-minded mission to make sustainability a part of how we innovate, build our brands, and build our culture.

That's what I've been doing for the past ten years. In 2019, I was named the first chief sustainability officer at P&G, and I'm now a part of the executive committee. It shows how sustainability has become a priority at P&G.

Pagitsas: Where does sustainability sit within the organization?

Helias: It originally sat within our biggest business, household care. Then we established a global chief sustainability officer role that is connected to all our businesses. When I was officially named chief sustainability officer, I made sure my position was anchored in the business. I always reported to one of the business leaders with the biggest business unit. That has worked very well to achieve my goal, which is to build sustainability into our business.

Pagitsas: Why was sitting within the business important?

Helias: It has to do with the shift of moving from CSR [corporate social responsibility] to sustainability. CSR is the old way of thinking and is centered on "cause-related programs," such as philanthropy, grant-making, and employee volunteerism. This isn't bad, but it's not sustainable. CSR is a department, and sustainability is a business strategy. CSR is a separate track and what we call "bolted on" versus being "built in." Today, we know that you have to change the way you do business if you want to stay in business. That's why sustainability is a business strategy and sits within the business.

When I explain sustainability's rise and role in a company to people, I often use the analogy of digital marketing. Fifteen years ago, digital marketing was a separate discipline. You would develop your marketing campaign, and you would build your TV advertising, and then, if there was some budget and time left, you would say, "Oh, what do we do online?" Today, our overall mix continues to move to include a growing share of digital and OTT [over-the-top] streaming.

This is where we want sustainability, and it's moving in that direction. It's not yet there fully. But if you ask me five years from now, sustainability will be built into the core business strategy, and my job will be made redundant.

Pagitsas: What do you do every day to move sustainability forward and embed it within your company?

Helias: What I do every day has dramatically changed in the past ten years. For the first two to three years, it was all about applying the tactics of change management. I am an executive coach on the side, which has helped me because my work back then was about understanding the deep motivation that our leaders had to do or not to do sustainability. Sometimes it was, "I need to do something about it." Other times it was, "Okay, give me something that can be a differentiator for my business."

I would say, "Any motivation, I'll take it," and I started working with those I called my "one percent." Those were the people who were already convinced about the value of sustainability. I said, "I'm going to start with this coalition of the willing versus using my limited capacity to convince the skeptics." Now there are fewer skeptics, but at the time, there were many. They could not see how sustainability could be part of their brand value proposition. I said, "If I can do things with those one percent, then I can start a movement and reach a tipping point." I worked with my one-percent leaders, and I started experimenting.

Pagitsas: What are some examples of working with the one percent?

Helias: The first thing we did was a partnership with TerraCycle to recycle some of our hard-to-recycle packaging. TerraCycle is a social enterprise on a mission to eliminate the idea of waste. We partnered on the air-care business, Febreze. We developed a powerful program where people were invited to bring their empty packaging back, and they will be turned into playgrounds for schools. That was very, very novel. That's how we started. We had many experiments like this. Ten years later, our collaboration with the Organizing Committee of the Tokyo 2020 Olympic and Paralympic Games through the Everyone's Podium Project delivered the first-ever medal podiums manufactured from recycled plastics collected from the public and recovered from the ocean.

But the true tipping point was the launch of the first Head & Shoulders plastic bottle made with plastic collected on French beaches. At P&G, brand-building and innovation are our superpowers. So, what was powerful with this change was that it was a brand-led idea. It was developed by my colleague Lisa Jennings, who oversaw the Head & Shoulders brand franchise globally. To create consumer awareness of the importance of recycling, they changed the brand's visual equity, which is not something we often do at P&G, from a white and blue bottle to gray because it was made of mixed-plastic. That was a very strong statement.

We announced the new bottle in Davos at the World Economic Forum in January 2017. It was one of that year's highlights in Davos, but the most exciting for me was the impact it had internally. The picture of the press conference with me holding the beach-plastic recycled gray bottle was broadly shared and liked. It was like saying, "We are now announcing to the world what we are doing, and we are doing it through our brands."

After this announcement, many of our internal brand builders started to pay attention to sustainability. That was interesting because those who used to ask for the business case as a cover for not participating were now fully on board. I would say, "But you know that the beach-plastic recycled bottle costs forty times more than a regular bottle." And they would say, "I don't mind. It's such a great brand idea. That's what I want for my business." Those experiments and those pilots went a long way to building this famous business case. Experimenting with brands is a very powerful strategy. That's a very bottom-up strategy. Once you've done this sufficiently, then you start doing the top-down.

I did the top-down strategy much later. This is where you make it systemic and ask your CEO to mandate it and make it part of all your systems. This is where we created Brand 2030, which is our new brand-building framework that empowers our marketers to make sustainability an integral part of their brand strategy and activation. We have set a goal that 100 percent of our leadership brands—twenty brands that represent more than eighty percent of our sales and lead their category—will enable and inspire responsible consumption.

P&G's growth model is based on what we call the "five vectors of superiority." To grow your brand, you need to have a superior product, superior packaging, superior communication, superior retail execution, and superior consumer and customer value. Our Brand 2030 framework builds on this and goes beyond it by asking brands to define their North Star—a measurable social or environmental commitment that fits with the brand equity and is brought to life over time with tangible acts. We are also broadening the concept of superiority by making sustainability a key part of it. This is what our consumers expect.

Pagitsas: What were the motivators that moved the other ninety-nine percent of P&G internal partners to engage in sustainability? Was it ten different reasons, or was it one?

Helias: It was a combination of seeing sustainability pilots build the business and growing consumer expectations. Nine out of ten people say they have a more positive image of a company when it supports social or environmental causes, and half say they make purchase decisions based on shared beliefs with the brand.

The Ariel Cold Wash story that I created is still one of the most powerful sustainability and brand stories. We delivered superior performance in a laundry detergent that takes stains out using cold water, decreased the carbon footprint, and grew the business. This is a win-win-win, all over. You can grow your business and do good for the planet. What's not to like in that? This year, we are launching the next generation of the Ariel Cold Wash innovation and communication campaign called "Every Degree Makes a Difference," and a similar one for Tide called "Cold Callers."

So, with increasing internal successes and growing external expectations, the question now is not, "Should we do it?" It is, "How do we do it?" and "How fast can we do it?" We've moved from building the business case for sustainability to the playbook to execute a sustainability strategy with excellence so that we grow our business and our positive impact at the same time.

Pagitsas: You talked about three very important stakeholders for a global company in the consumer-packaged goods industry—the consumer, the retailer, and the investor. Are they asking for different things from P&G or the same?

Helias: Generally speaking, they're asking for the same thing, which is "We want more, and we want it faster." When you unpack that, you see that they have some differences. If you look at consumers, the number-one thing consumers ask us to resolve is packaging because this is the most visible. Today consumers come back from grocery shopping, and the first thing they see is extra packaging that they have to throw away. Then the second thing is enabling them to reduce their energy and water consumption at home. Last, ingredient safety is obviously a prime expectation that P&G takes very seriously, having hundreds of scientists in our safety and regulatory area for the company.

The general public's awareness has increased during the lockdown period. With people spending more time at home, they have also realized how much more waste they were producing, and how much more water and energy they were using. And they are asking brands to help them deal with that.

Retailers are basically following consumer preferences, but they're also going ahead of them. For instance, certain retailers have said, "If your packaging doesn't have a certain amount of post-consumer recycled plastic and is not recyclable, your product can't be on our shelf." That can be motivating for sustainability across the industry! For investors, the focus is on climate and recognizing that the next decade is our window of opportunity to mitigate climate-related risks.

Climate is extremely relevant for P&G because seventy percent of our products are used with warm water. While you shave, do your dishes, wash your hair, wash your clothes, all this is heated water. We have been reducing our emissions for decades, notably by formulating our products to enable consumers to use less water and colder temperatures with no performance trade-offs. Homes and the products produced to be used in them globally are responsible for more than a third of all the emissions.

Pagitsas: How do you innovate at P&G to enable responsible consumption?

Helias: We start with science because we want to make sure we will innovate where we can have the greatest impact. This is why life cycle analysis is the backbone of our sustainable innovation strategy. Once we understand this, we work on the aspects within our control, like reducing the amount of virgin plastic in our packaging—our goal is to reduce it by fifty percent by 2030—or using sustainably sourced material.

Over eighty percent of our carbon footprint is related to people using our products with heated water. Although this is not entirely within our control, we believe we are responsible for enabling people to use our products more responsibly, like we are doing with our laundry detergent cold wash innovation and communication. We develop products that work better in cold water so that people can use lower temperatures. That is a winning partnership.

Another example of how our product innovation enables responsible consumption is Cascade, a major dishwasher soap product in the United States. Cascade saves water and energy if you run your dishwasher versus doing your dishes by hand. This is because when you use the sink tap, you use four gallons of water every two minutes. In contrast, four gallons of water is what a dishwasher running a full load uses. Do the math. If you are washing more than eight dishes, you're better off running the dishwasher.

When people learn that, they say, "I didn't use my dishwasher because I thought I would waste too much energy and water." This is what we call a transformative insight that drives sustainable habit change. But you can only do that if a product like Cascade performs well enough to allow consumers to skip the pre-rinse of dishes before loading them into the dishwasher. The ultimate in enabling responsible consumption is what we call "making sustainability irresistible." You make the sustainable options so desirable that even if people are not interested in sustainability, they go for it. That is the concept that is behind our 50 Liter Home program.

Pagitsas: What would you tell sustainability leaders who are struggling with engaging their C-suite on this topic?

Helias: I would tell them two things. The first is to make sure you bring "the outside in" to help them navigate this complex and dynamic environment and understand best practices. This is why I created an external advisory board five years ago. I gathered the finest thought-leaders and practitioners. You know when you ask the question, "If I could have dinner with anyone, who would be on my list?" That is what I did for my board. I wanted the best sustainability leaders around the table with the C-suite to inspire, guide, and challenge us. Those meetings—chaired by our CEO—are amazing. Straight talk, education, and inspiration followed by action. We take the feedback very seriously. It is that group that helped us develop our Ambition 2030 roadmap.

The second thing I would advise my fellow sustainability leaders to do is build strong partnerships with the C-suite—beyond the CEO. For me, it started with the chief brand officer, Marc Pritchard. When I started the job, he was not particularly engaged on the environmental front. I felt that he saw it a bit as "Virginie's thing" until he received an invitation to speak at a major sustainability conference.

Marc is an amazing man. When he commits to something, he rolls up his sleeves and delivers! For three months, he went to school on sustainability as he was preparing his talk, which ended up being the highlight of the conference. It was amazing. In the process, he learned about the topic. Like everyone, the more you learn about sustainability, the more you're passionate. This started our partnership which has developed ever since and has been game-changing. It has been a key enabler to engage our brand builders. It has given birth to what he calls the reinvention of brand-building—Brand 2030.

Pagitsas: Why was it important to have Marc tell the story rather than you at the conference?

Helias: We are a company of brands. I always say if sustainability is not on the brand agenda, then it's not on the agenda, period. It has to start with the brand agenda. We touch five billion people through our brands globally every day. Even small steps multiplied by five billion will add up to a significant collective impact. Through that partnership with Marc, we accelerated the integration of sustainability in brand-building and developed our Brand 2030 framework. From that moment on, I decided that every year I would internally develop another game-changing partnership.

Pagitsas: With whom will your next C-suite partnership be?

Helias: We have a new chief research, development, and innovation officer. This is another critical partnership. We have agreed that our lighthouse will be to make every innovation a sustainable innovation. This will enable people to live a more sustainable life by using our brands. Over the past few months, another partnership has become a business imperative, which is between the CSO and the CFO. As more and more investors are making sustainability and environmental, social, and governance key investment criteria, you can imagine how this can be transformational!

Pagitsas: How do you balance setting challenging objectives while still making sure you don't lose the organization's engagement?

Helias: This is a fine balance. You want to see challenging targets because that is what the world needs, but you also want to make sure you can operationalize them in a way that builds our business and reduces our footprint. Sometimes, I call myself the "chief agitating officer" because I consider my job as pushing the lines and constantly raising the bar, but you need to do it in a way that makes sense for the business.

My role is to synthesize the different, often diverging, perspectives from all our stakeholders—consumers, regulators, NGOs, and investors—and offer frameworks for our business leaders to make the most informed decisions that balance economic, social, and environmental considerations. This is where my thirty-year tenure gives me such an advantage. Because I know the business and the culture, I know how things get done. I always have so much respect for CSOs who are hired externally. I could not have done it at P&G.

Pagitsas: How do you know which issues to tackle when business, environmental, and social issues are dynamic?

Helias: We conduct a sustainability prioritization exercise by interviewing internal and external stakeholders, including investors, NGOs, peers, and suppliers. Among a list of topics, we ask them which they believe are strategic for our business, and they see us having the greatest responsibility to tackle. That gives us a matrix of priorities. It's interesting to see the issues that they believe are very important externally and what we may underrate internally or the other way around, so it triggers a discussion. For example, some external stakeholders did not see water as the highest priority, while we know how critical access to water is for us since most of our products need water to be used. With that, we can understand what the external demands are. Then we put our business lens and say, "How are we going to prioritize and operationalize?"

Pagitsas: Would you say that every company has issues important to their business? And knowing about them through systemic evaluation is the best way to develop a strategy to mitigate them?

Helias: Absolutely. Prioritization has been the starting point for developing our Ambition 2030 goals. We use the lens of science to make sure that we work where we can make the biggest difference. It has allowed us to develop the roadmap for 2030.

Pagitsas: What is your favorite book to recommend to new CSOs?

Helias: *Switch: How to Change Things When Change Is Hard* [by Chip and Dan Heath (Crown Business, 2010)]. This is the definition of a CSO job, and it has been a very enabling tool for me on the journey.

Pagitsas: If you were sitting in a room with a CEO and the board of another Fortune 500 company that was starting to develop its sustainability strategy, what would you tell them?

Helias: I would tell them three things. One, know where your greatest impacts are. Science will tell you. Listen to experts. Two, don't do it alone. No one company will be able to meet the planetary challenges we face at the speed and scale we need. Join cross-sector, public-private partnerships, and coalitions to drive impact at scale. And three, be a role model for courageous leadership. We need more of this, and we need it fast. I would urge them to be an activist, an artist of change, an alchemist, an impactful leader, a trailblazer, and a lasting trendsetter, and make sustainability irresistible before climate change becomes irreversible!

Marissa (Pagnani) McGowan

Chief Sustainability Officer
PVH Corp.

Through October 2021, Marissa (Pagnani) McGowan served as the chief sustainability officer (CSO) at PVH Corp., one of the largest global apparel companies. PVH has a portfolio of brands that includes Tommy Hilfiger and Calvin Klein. The company had $7.1 billion in revenue in 2020. Marissa currently serves as chief sustainability officer at L'Oréal, NA.

As chief sustainability officer at PVH, Marissa led a global team responsible for the strategy and execution of PVH's corporate responsibility mission, Forward Fashion. Marissa and her team designed and implemented programs aligned to Forward Fashion's priorities to uphold human rights and drive environmental sustainability, including capability building, stakeholder engagement, and non-financial sustainability reporting.

© Chrissa Pagitsas 2022
C. Pagitsas, *Chief Sustainability Officers At Work*,
https://doi.org/10.1007/978-1-4842-7866-6_5

Marissa joined PVH when the Warnaco Group, her previous employer and the largest licensee of Calvin Klein, was acquired. At Warnaco Group, she was the in-house legal counsel leading legal teams and work in Europe, Asia, and Latin America, and running the company's human rights and chemicals management programs. She began her career as a lawyer at Skadden, Arps, Slate, Meagher & Flom LLP focused on international mergers and acquisitions and corporate compliance.

Marissa served on the Global Fashion Agenda steering committee and the boards of the Accord on Fire and Building Safety in Bangladesh, the Industry Association, and the Fair Labor Association.

Marissa has been named among the Top 50 ESG People of the Year, the Responsible 100, and Environmental Leader 75 honorees. She is also a member of World 50's Sustainability 50. While Marissa was chief sustainability officer at PVH, the company ranked number 12 in Barron's *100 Most Sustainable Companies and number 16 out of 400 in* Newsweek's *America's Most Responsible Companies.*

Chrissa Pagitsas: Many readers may know of the brand names Tommy Hilfiger and Calvin Klein but may not have heard of PVH. What is PVH and its relationship to those brands?

Marissa McGowan: PVH was founded 140 years ago, making it one of the oldest and largest apparel and fashion companies. The company is known best for our brands such as Tommy Hilfiger, Calvin Klein, and Warner's. With over 33,000 associates worldwide, we operate more than 6,000 retail stores and have a presence in forty countries globally.

Pagitsas: What is the history of corporate responsibility [CR] at PVH? And what is the scope of CR today?

McGowan: PVH has been active in the corporate responsibility space for three decades. We were one of the first companies to have a factory human rights code of conduct, established in 1991, meaning a set of standards based on the International Labour Organization's [ILO] conventions for how a factory upholds human rights. We're a founding member of the Fair Labor Association and one of the first companies to have a committee of our independent board of directors dedicated to corporate responsibility created over ten years ago. We've also been publishing CR reports for thirteen years.

While I've been with the brands for fourteen years, I started in the sustainability role about eight years ago. When I began in the role, it was a different program. We had a vision to evolve a well-regarded factory audit program focused on human rights into something more comprehensive, spanning the entire apparel value chain, from the cotton farm to the mill turning yarn into fabric, to the factory where we're cutting and sewing the T-shirt, through to the consumer use of the T-shirt, and the end of life for that T-shirt. We also wanted to encompass environmental sustainability, social impact, centralized data, and best-in-class reporting and transparency practices to complement our human rights work.

Today, our CR strategy reflects that vision from eight years ago. The work is built around Forward Fashion, which establishes a roadmap to deliver PVH's purpose of "powering brands that drive fashion forward—for good." The strategy focuses on three strategic pillars—reducing negative impacts to zero, increasing positive impacts to 100 percent, and improving over one million lives across its value chain. These pillars include fifteen priorities and time-bound targets with focus areas ranging from eliminating carbon emissions and hazardous chemicals, innovating for circularity and sourcing ethically, amplifying worker voices and living wages, to empowering women and fostering inclusion and diversity.

Pagitsas: That is a big expansion of the strategy! Was it overnight? Or was it a series of steps to get to the Forward Fashion strategy? What drove this shift?

McGowan: It was a gradual set of trust-building steps built on the foundation of a complete partnership mindset. A partnership mindset means acknowledging that a siloed team cannot drive the change we want to make in the company, and a company alone cannot drive the change we want to see in the industry. This shift happened because we invested time and energy into building relationships within the company with all aspects of the business and operations, as well as committing resources to engage with civil society and take active roles in industry initiatives. Partnership and collaboration are hard work. It's very different to say you're a partner and to act like a partner.

Pagitsas: What is a successful example of you and your team using the partnership mindset?

McGowan: Here are two examples relating to how we built trust and were able to expand our impact. First, in the early days of our environmental work, we had energy information for sixteen owned and operated sites—offices, warehouses, retail stores—for our greenhouse gas footprint. We now have energy and water use data from approximately 2,600 owned and operated sites and over 600 factories in our supply chain collected annually. How did we do that? We started by learning who holds the relevant data, such as electricity usage on our offices, distribution centers, and retail stores, and then building relationships with those teams.

We then said, "Let's make this a two-way dialogue with you, our in-house partners. We will share our objective of measuring and lowering our greenhouse gas emissions, which can have business benefits." For example, more efficient energy use means lower energy bills. We said, "If you can participate by sharing some of your data, we will come back to you with information about how your stores, offices or warehouses perform from an energy efficiency perspective against others in our portfolio and as compared to other similarly situated facilities outside our company. We can share best practices and tips for increasing our efficiency and would like to hear from you what problems you are trying to solve and what has and hasn't worked in the past." Rather than demanding data without context or making it a simple data pulling exercise, we created an environment where it was a joint effort in which we each had roles and opportunities to contribute positively toward a shared objective.

Another example is in our supply chain, which includes about 2,500 facilities globally where we source both finished goods and fabric directly and through licensees. Those facilities employ about a million people, so there's quite a big human impact there if we do things well. When I started, we were conducting audits at around 450 factories. By the next year, we were conducting audits at 2,500 factories. Why the increase? This speaks to the importance of building relationships with the different supply chain teams around the company.

We worked with the supply chain teams to understand their sourcing footprints and have a two-way exchange. We explained the importance of knowing where we produce and how our programs helped manage risk through audits and add to positive impacts in our supply chain communities through programs to support workers and increase environmental efficiency, for example. We created policies with clear touchpoints, including reflecting together on audit data to ensure we all understood what needed to happen to improve and could speak to our suppliers with one voice, whether we were CR associates or sourcing associates.

Pagitsas: What were the outcomes of these conversations?

McGowan: Ultimately, these dialogues took us to a place where we gained respect and credibility internally for being business partners who wanted to work together. Once we had robust factory management and greenhouse gas reporting functions in place, there was an openness to allow our team to work on more progressive human rights issues and expand into new topical areas like chemical management and water stewardship. These issues are highly relevant for the fashion and apparel industry because we use a lot of water, energy, and chemicals in the process of growing cotton, spinning yarn, dyeing fabric, etc. Continuing to execute successfully as we expanded the scope of our work culminated in 2016 with the publication of ten commitments around human rights and environmental sustainability.

This first set of public CR commitments was not SMART [specific, measurable, attainable, relevant, time-bound] targets like we have today, but they were the first public stake in the ground. There was a lot of work happening behind the scenes, but before this, we weren't necessarily comfortable talking about it. It's a testament to the company that it pushed itself to be more transparent and to hold itself accountable. We kept building the momentum up with those ten and resulted in where we are today, which is Forward Fashion with a set of fifteen priorities, and I believe, a new level of ambition and transparency that we're hoping to deliver to the industry.

Pagitsas: While the initial work grew organically under your leadership and had a bottoms-up implementation, it sounds like today's full strategy is an enterprise strategy with top-down support from PVH's leadership.

McGowan: Yes, both our former CEO and our current CEO are very committed to the work. By having a board-level committee, we have top-level visibility and oversight. We also do a ton of benchmarking against our competitive set, which includes more progressive competitors, which has helped with strategy- and goal-setting. We say, "Sometimes we're pulled, sometimes we pull." That's the importance of recognizing your role as an industry player and a player in a larger ecosystem, which includes competitors and civil society.

Pagitsas: Speaking of senior leadership, how do you engage with the board, and where does your team sit within PVH? It's different in every company.

McGowan: I have very regular and good access to senior leadership because we have a board committee that meets quarterly to discuss our sustainability strategy. As sustainability touches all parts of the business, my role could sit anywhere. I'm currently reporting to our chief supply chain officer and our chief people officer. Over the years, I've reported to the chief risk officer, the chief operating officer, the chief finance officer, and the general counsel.

Pagitsas: What's an example of partnering with the board on PVH's sustainability strategy?

McGowan: This spring, we held small deep-dive sessions with the board on the next chapter of our sustainability work—embedding the work into the business versus having a central team do the work. We had an opportunity to engage with our board members who are incredibly experienced individuals from all different industries. They come to the conversations with different lenses. It was a great experience to work with them on the next chapter of sustainability focus for the company. We thought about how do we get the most out of this time with them? We were encouraged by our CEO to provide them with pre-reads and give a short presentation, but to let them ask the questions to get the most out of their experience. We discussed questions such as, "If our goal is to get the company aligned behind the principles of circular economy, what would resonate with you as a board member? Is this an efficiency and business process discussion? Is this a natural resource and climate change discussion? Or is this a discussion about what's 'cool and shiny'?"

The outcome of the conversation was an endorsement of our approach to focus the next phase of work on the consumer, our business operations, and what we are calling "next-generation stakeholder" groups. Within each of those areas, we were able to hear where the board thought we should focus. For example, with the consumer, should our focus be on e-commerce or brick-and-mortar stores? Should our product focus be on improving all of our products slightly or making our best-sellers the most sustainable? Coming out of those conversations, we knew we had board support and input into our work and guidance to discuss with management on where to focus resources.

Pagitsas: Is there a push from investors as well as the board to engage in sustainability?

McGowan: Over the past years, socially responsible investors and some of our European investors have asked us about our sustainability strategy and our work on key issue areas such as setting and paying a living wage to the workers in our supply chain. Yet, we're now seeing sustainability come up much more with mainstream investors. When I say "much more," we're not talking about every conversation, but it is more than we were hearing in the past.

What's super interesting about the investor conversation is that you might think the questions are going to be around climate change, our human rights program, or the living wage. Instead, I find the conversations more focused on systems and how we are building resiliency and the mechanisms for long-term value. Maybe I'm reading too much into the way the questions are being asked, but they seem less topic-specific and more about how we are thinking about these issues systematically, managing the risk, and putting the systems in place to address the issues. That has surprised me. It's not to say we don't still get investor surveys and inquiries with very specific questions about our goals and how we're progressing against them, but conversations with investors get interesting when the investment community and business leaders connect about companies that are focused on sustainability with being either more efficient or more resilient.

Pagitsas: What are some examples where the business has been more efficient or resilient as a result of its CR strategy?

McGowan: When COVID-19 first hit in Asia, there were obvious effects like factories closing, which raised questions about worker health and safety, payment to workers, and the like. As we got into the early months of COVID-19, we experienced the ripple effects, very fast, very furious from a supply chain perspective. The impacts were not just that a factory was closing down. It could be that the factory was only at partial capacity, or you couldn't get an auditor into a factory because no visitors were allowed. As the COVID-19 pandemic spread to different countries and we mitigated against an impact in one country, something would pop up in another. You'd mitigate against that, and something would pop up again. We were onboarding new factories at rates we had never experienced before so that we could have business continuity.

We were also trying to work with suppliers with whom we had pre-existing relationships to deliver PPE [personal protective equipment] and other supplies. Some companies had a tough time finding supplies. However, our investment in best-in-class suppliers, who had compliance at the top of their list and invested in their workforce and their communities, meant we were able to keep goods flowing from one supplier to another. As one factory went offline, another one came online because of the relationships we had built over the years. That is an example of supply chain resiliency. This is proof of the importance of strong supplier-level partnerships, and it came to action.

The importance of the time and effort we put into working with civil society became abundantly clear during COVID-19, even though sometimes it can be looked at as lip service or something we do for PR [public relations]. Once it became apparent how serious the pandemic was, my team and I were on the phone with the ILO's Better Work team and a few of our peers at other companies. Within about two weeks, including many nights, we had an agreement with the international unions whereby brands would pay for their finished goods and work-in-progress already at factories.

We worked together on leveraging financing mechanisms to make sure workers got paid, and we would commit, over the longer term, to work on safety nets, social protections, and social insurance programs. That agreement, called the COVID-19: Action in the Global Garment Industry, came together because we have invested in those partnerships. Again, that's the big difference between sending money or a check and building programs on the ground together and having the phone numbers to call when you need to.

Pagitsas: A takeaway is that you made an investment in partnerships before you needed them, essentially investing before you knew the dividend, whether the ILO or the suppliers.

McGowan: Absolutely.

Pagitsas: It sounds like the investment in stakeholder engagement, which is a principle of sustainability, had a financial benefit. Does every sustainability project need to have a financial return on investment [ROI]?

McGowan: I have long struggled with that concept for multiple reasons. There are obvious examples such as the link between lowering energy use and, therefore, energy bills. You can eco-size your labeling and packaging, which means spending less on packaging. Those are nice tidy examples of ROI. However, other areas such as the stakeholder engagement from the last example are not as easy to quantify. No matter how good the statistics are, there may be competing statistics in a different direction. Isolating sustainability as the cause or effect of one of those things is incredibly difficult.

Pagitsas: How do you know when to make an investment in a sustainability project with an unknown financial ROI? How do you convince others to support it?

McGowan: Sustainability has to make sense for the business, whether the goal is increasing associate engagement or investor differentiation or reaching out to a different consumer segment. However, knowing your audience is also important. There are audiences who want to make a difference. They want to use our platform to drive some type of positive change. It's a hearts and minds conversation around making an adjustment to the way we source, the materials we choose, or the impact we have on a community. There are others who are very focused on business efficiency and hate waste in the system. Whether that's eco-sizing a label on your clothes or looking at a business process and seeing an inefficiency, we could see a lot of that because of the impacts down the road on human rights and environmental practices.

Sometimes it's a hearts and minds conversation. Sometimes it's an efficiency conversation. Sometimes it's an employee morale conversation. The more clued in you are to the agenda and goals of the people you're working with and for, the better you can be at that and the more you know them because that may change over time. You may have a hearts and minds person who is very focused on efficiency that particular day. [laughs]

Pagitsas: Do you think that there is an outsized responsibility for chief sustainability officers to be able to have hearts and minds conversations that is not expected of other C-suite peers such as the chief marketing officer or chief technology officer?

McGowan: Yes, I do.

Pagitsas: Why is that? It's a unique job requirement in comparison to the other roles in the C-suite.

McGowan: Yes. It's not changed, but I think it's changing. Historically, sustainability was tied to philanthropy, so there is a bit of an "aid versus trade" conversation. Is this just a contribution to the social good? Is this a contribution to a social good that also has a business benefit, or is this really a business strategy that also has a social good? You have this spectrum of what role the work is really playing.

Another challenge of this role for CSOs and the scope of sustainability for companies is that nothing has been standardized, so we don't have great mechanisms to measure ourselves against others or to be differentiated in the regulatory, investor, or even consumer environment. Until this occurs, it is up to us to contribute to credible efforts to standardize and be early adopters of what is out there. When key stakeholders, such as consumers, associates, investors, financial institutions, and regulators, begin really rewarding the companies making the biggest positive impacts, the virtuous cycle takes over, the impact we can make is exponential.

Pagitsas: Today, PVH does have ambitious corporate responsibility targets and a strategy to deliver on them. However, every company, particularly large companies, has critics. How do you view those critics?

McGowan: A very wise woman once told me there are three types of critics—dolphins, killer whales, and white sharks. The dolphins are the critics that are there to push you, but they're going to work with you. They're very excited to stay together and work together. They're going to push you a little bit but not call you out. Then you've got your sharks for whom, no matter what you do, that's not going to be good enough. They are going to eat you no matter what. There's very little you could do to appease them. You then have the killer whales. They push you harder than a dolphin will, but they do it in a way that's inspiring and motivating.

We see a huge role for our critics in terms of keeping us moving forward and accountable. I use this example to differentiate our critics and believe you have to approach them differently as well. If you can find those killer whales, it's quite exciting to work with them.

Pagitsas: Who's a great killer whale that inspires you and pushes you to do better?

McGowan: The International Labour Organization's Better Work program has historically partnered with the apparel and footwear industry on human rights issues, so they are my human rights killer whale. Some of my most interesting and provocative conversations are with them. They push hard. They have high standards. Because of their tripartite nature, unions and governments are always in the conversation, so you're constantly pressuring yourself to think about how do we better leverage the ecosystem. Because you're working under international conventions, there's no, "Oh, let's just lower the bar a little." That's not how it works. There have been a number of times where I've wanted to work on something with them, and they've said, "Okay, then this is what we want to see from you."

To go back to my earlier story about the COVID-19: Action in the Global Garment Industry agreement, there was a lot of discussion at the time about committing to longer-term social insurance and safety nets. But it was a non-negotiable for the ILO and its Better Work team to be engaged. And they were the lynchpin to securing an agreement that involved both business and the international unions.

World Wildlife Fund [WWF] is a good example of a killer whale in the environmental space. In order to work on water stewardship projects with them, we had to make SMART commitments in terms of our chemical management, water use, and carbon footprint. They're a great partner, and they're great implementers. With WWF, we worked on our global water risk assessment and footprint and eventually agreed to stewardship projects in key water basins for us, including Lake Taihu in China, Lake Hawassa [also spelled Awasa or Awassa] in Ethiopia, the Kaveri [also spelled Cauvery] river in India, and the Mekong basin in Vietnam.

There's a give and take that comes with partnerships like these. Because you have mutual respect, you also want to keep pushing the boundaries together.

Pagitsas: I agree. The right partners are invaluable. What are the biggest critiques, whether from dolphins, killer whales, or sharks?

McGowan: We're seeing an increased conversation around consumption more broadly. The focus is on the waste in the fashion and apparel system and what is going into the landfill, as linked to the limitation on natural resources and the importance of moving to a circular economy. We do have a lot of work to do as an industry, but I'm excited to see what's resonating with the consumer today around things like claims of durability.

As a consumer, will you pay a little bit more because this is a more durable product that will last longer? Obviously, companies like Patagonia have been doing that for years. It's part of their ethos, they're a B Corp that has been certified by a third party as meeting certain social and environmental performance, public transparency, and legal accountability measures to balance profit and purpose. But we're seeing it much more in the mainstream.

Increased use of more sustainable materials is of growing importance, but the solution to the circular economy discussion gets very complicated quickly because we don't yet have the innovations or infrastructure to take back and recycle garments in a scalable way.

The biggest critics that I listen to the most are the ones saying the transformation is happening too slowly. To them, I keep going back to how can we differentiate and incentivize companies in a different way? Is there ever going to be some standardized measurement? Vis-à-vis the consumer, we have been working through the Sustainable Apparel Coalition's Higg Index on the equivalent of a food label on a garment. Instead of calories and vitamin count, it tells you the energy, chemical, and water use, the sustainability of the material choices, and the human rights and social impacts of manufacturing and distribution. It includes what a consumer can do to help.

How do we make that super simple so that our consumers can make an informed choice? It can't be that every company has its own system. It has to be one system. The same thing can be said from an investment and a regulatory perspective. What are those measurement tools that are going to signal that we're doing all this work, so we're going to get a better financing rate? Or that we have an advantage where our products can meet progressive environmental regulations in the European Union ahead of our competitors? I always go back to the ideal state of some standardization in data reporting and customer communication, but that seems a long way off and maybe like a bit of a dream.

Pagitsas: Given these barriers, if there were one change that you'd make in the fashion and textile industry because it's so important to the environment and to society, what would it be?

McGowan: Full input such as components, materials, etc., and production traceability. Imagine a small tracker on every component of a garment, as well as the finished garment, telling us where the inputs came from and where they were assembled. Where we see the greatest problems and the greatest opportunities are where things are opaque. They are opaque in many places along a very long value chain. This applies to environmental issues just as much to human rights issues.

I don't believe the apparel and fashion companies that tell you they're doing tons on environmental issues but don't know where their fabric mills are. The fabric mills are where the greatest negative environmental impact is. You'll see that very often. A company will produce a list of factories, which they contract with for the finished garment. When you drill into that list and ask where the raw materials are coming from and where the fabric came from, they say, "Oh, an agent does that for us," or "We bought this from the factory as a package— they source the raw materials for us." Then how do you know that you're making a difference in your environmental commitments? What is your baseline and how did you set it? The factory is just one step back in the value chain. You have to keep going back through the fabric mills, the spinners, the ginners, and the cotton fields. There are environmental impacts throughout, not to mention the labor issues at all levels, with respect to the informal economy in particular. Having a mechanism to be able to trace and track would be incredible.

Pagitsas: Doesn't this technology exist today?

McGowan: Apparently, there are tons of them. We talk to people all the time, and they were going to be underpinned by all sorts of blockchain technologies, but I have yet to see a company that's been able to scale it for the fashion and apparel industry.

When you look at all the inputs, you're going to need to trace your button, the zipper, and the rest of the materials from the factory, the mill, the spinner, and the component processors. That's the future. The sooner we can get there, the better we're going to understand where our problems are. The better we understand them, the sooner we can make commitments, and the sooner we make commitments, we can start to work on them and hopefully be held accountable by the killer whales. [laughs]

Pagitsas: Nice tie into those killer whales! Let's talk about the UN Sustainable Development Goals [SDGs]. Why is SDG alignment important? How does PVH's sustainability work align with the SDGs?

McGowan: The SDGs are brilliant. They give us a common language to speak across the industry—from industry to government to civil society, and they show the interconnectedness of the issues we are working on.

It's fun each year when we do our CR report to map each story to multiple SDGs. You see a story and may say, "Oh, this is a water story only." But it's also a story about decent work, *and* it's a story about resiliency, *and* it's a story about climate. You start to see how connected these environmental, social, and economic issues are. For creating the common language and identifying the interconnectivity, I applaud the efforts of those who developed and maintain the SDGs. If we're going into a country for the first time as a sourcing destination in a big way, it's interesting to see what is that country's plan against the SDGs and how we can fit into that by bringing jobs or another benefit aligned to the SDGs.

We did an alignment analysis with the UN SDGs when we went into sub-Saharan Africa. For example, in Ethiopia, we focused on building industry and employment for a young population, which aligns with SDG 8, Decent Work and Economic Growth. We also focused on sustaining fresh water, which aligns with SDG 6, Clean Water and Sanitation, and investing in clean energy, which aligns with SDG 7, Affordable and Clean Energy. As a result of this mapping, we worked with WWF on a water stewardship project there, as well as ensuring the factories we sourced from used water systems that recycled as much water as possible.

Pagitsas: You are well-versed in the SDGs and know how to navigate them. What would you tell a sustainability leader jumping into aligning their business' work to them for the first time?

McGowan: Someone asked me right when the SDGs came out, "How did you bring this to your CEO?" I said, "Under no circumstance would I ever bring this giant framework to him." My job is to translate it, give an overview and relate it to work he is already aware of. So, I'd focus that sustainability leader on identifying how they're already doing on certain SDG goals. Here's where they can push the boundaries, and here's how they can ease into it, instead of saying to their CEO, "Here they are, and here's what we need to do."

Pagitsas: Let's circle to your educational and professional journey to PVH. What experiences influenced you most?

McGowan: My professional journey is first grounded in my experience going to a Jesuit college and graduate school with a large international student body, Georgetown University. The Jesuit experience focused on the principles of service for others and the belief that unto those who much is given, much is expected. The experience with the Georgetown Law Center's connection to civil society, government, and firm life was critical to my career because it set me up on the path and with the relationships that got me to where I am today.

After school, I went to practice law at Skadden, Arps, Slate, Meagher & Flom LLP. I have three key takeaways from my experience there. First is the importance of being trained properly. I was taught to know my audience, pay attention to detail, have grace under pressure, be rigorous within time constraints, and use judgment. All these things seem obvious, but I have found being deliberate about them has greatly served me throughout my career. My second key takeaway is the role of mentors. I owe so much to my mentors, Tom Kennedy, Rick Grossman, Alan Myers, Fred Pagnani, and Keith Pagnani, because they invested in me, giving me solid and direct feedback, which at times was hard, but in retrospect, I'm grateful for it.

The third takeaway is the role of the quarterback in the legal arena but also corporate responsibility. As a mergers and acquisitions lawyer, I was taught to work with many of the other specialty legal teams such as corporate finance, employment, intellectual property to capture and reconcile their perspectives into an agreement. As chief sustainability officer, there is a similar quarterback role. Each lever you pull impacts the other. You have to be thoughtful about how things fit together and impact the other pieces of the ecosystem.

Skadden took me to London, where a client, Warnaco, was buying the Calvin Klein jeans licenses. About a year later, the general counsel from Warnaco, who I knew from Skadden, called and said, "We've now got this international business. Would you like to set up a legal team in Europe and Asia?" During my Warnaco time, I was a sponge. I just wanted to learn the business and the operations, as well as the differences between the European, Asian, US, and Latin American businesses. I learned how to be a conduit between corporate and a region, which I think is a great learning experience, and to navigate cultural differences.

The biggest takeaway from the Warnaco experience was the type of lawyer and professional I wanted to be. I loved understanding what the business was trying to achieve and then seeing where there are roadblocks from a legal perspective. How can we navigate this in a way that's most responsible but also pushes the boundaries so that we can do things that advance the business? As chief sustainability officers, we're constantly playing in what I call "the gray." There's not necessarily a right or wrong answer. There's a balance of pros and cons. There are risks and rewards. There are unintended consequences. There are few things I find more rewarding than navigating challenges in a way that helps the business advance in the most responsible way. Few things are more rewarding.

Finally, I came to PVH because they bought Warnaco to reunite the house of Calvin Klein. It was an important professional experience to live through a company's acquisition and incredibly humbling. Mergers and acquisitions have a synergistic aspect, which may mean you have duplicative roles. I said goodbye to many colleagues and friends during that time. There was also a lot of opportunity in joining a new organization that had bold plans on human rights and the positive impact they wanted to make.

I joined PVH as a lawyer, and two of my first assignments were favorites from a career perspective. The first was joining the implementation team for the Accord on Fire and Building Safety in Bangladesh. The Rana Plaza building collapse [in Dhaka, Bangladesh, in 2013] was the greatest industrial disaster the apparel industry suffered. PVH did not source in Rana Plaza. A group of companies, including PVH, signed a legally binding agreement with the international trade unions, IndustriALL and UNI Global Union. NGOs [non-governmental organizations] were witness signatories and participants in the negotiations. We had this five-page agreement but needed to negotiate every aspect of turning the agreement into a fire, building, and structural safety inspectorate. I was lucky enough to be one of six company representatives sent to Geneva to negotiate everything from where to set up our legal entity to what data could be disclosed and to hiring our first executive director and fire safety officer.

The second assignment, in short, was looking into criminal anti-sodomy laws in countries where we were considering establishing sourcing operations. I believe the passion with which I approached those projects and the relationships I built through them made me a contender for the role of heading corporate responsibility when my predecessor moved on for personal reasons. And then I found myself in the job I always wanted but didn't know existed.

Pagitsas: If you had to sum up your personal leadership approach to strategic, top-level discussions on sustainability at a global, multibillion-dollar company, what would it be?

McGowan: I would say three key things. First, with all executives and everyone I work with, I approach with humility, listen before speaking, am prepared, and translate the work into something that's meaningful to either their agenda or their goals.

Second, I always try to bring relevant information, which they may not have access to or focus on. My team and I engaged in regular benchmarking against our competition and cross-sector sustainability leaders. We invested considerable time in building relationships with peers at our competitors and understanding the trajectory of trends, investments, and innovations in the sustainability space. Bringing relevant information to meetings helped leaders consider and take decisions on sustainability topics.

Finally, I tried to improve our team with every hire and invested heavily in growth and development. An incredibly smart and talented team, versed in both technical and soft skills, goes a long way in both delivering results and building credibility. I believe all those things have served me very well in terms of engaging on the sustainability topic with peers, colleagues and leadership. Luckily, I work with people who are excited to engage on the topic.

Pagitsas: What do you hope will change in the fashion and apparel industry as it further embeds sustainability principles into its operations?

McGowan: For the fashion and apparel industry, as with all industries, I hope first that sustainable business proves itself to be good business—moving from a mindset of needing to do well financially to do good to a mindset of sustainable business capturing the efficiency, resiliency, innovation, strong relationships, and trust that are essential to long-term value creation. Second, I hope for the genuine recognition that to address system-level challenges, there is an important role for business, government, and civil society to work together differently. It's the only way we will solve the major challenges ahead of us.

Jyoti Chopra

Chief People, Inclusion and Sustainability Officer
MGM Resorts International

Jyoti Chopra is senior vice president and chief people, inclusion, and sustainability officer for MGM Resorts International. MGM Resorts is a global gaming, hospitality and entertainment company with locations on the Las Vegas Strip, across the United States, and in Macau, with over $8 billion in revenue for the last twelve months ending September 30, 2021. Jyoti is responsible for leading human resources and social impact and sustainability and overseeing the MGM Resorts Foundation. She oversees environmental, social, and governance (ESG) reporting, is responsible for directing enterprise-wide human and social capital initiatives, and serves as a liaison to the board of directors' corporate social responsibility and human capital and compensation committees.

Prior to her role at MGM Resorts, she was senior vice president of Global Diversity & Inclusion and Human Resources Operations and Transformation at Pearson Plc. Earlier in her career, Jyoti served as the chief diversity officer and managing director of Global Citizenship and Sustainability for Bank of New York Mellon, as well as the global leader for communications and public relations at Deloitte Touche Tohmatsu Limited and as a managing director at Merrill Lynch & Co. Jyoti began her career covering human development and issues affecting women and children at the United Nations.

© Chrissa Pagitsas 2022
C. Pagitsas, *Chief Sustainability Officers At Work*,
https://doi.org/10.1007/978-1-4842-7866-6_6

Jyoti is a member of Toyota Motor North America's Diversity Advisory Board and serves as a diversity and inclusion advisor to the Spencer Stuart Global Leadership Team. She is a member of the board of directors at Schneider National and serves on the company's compensation committee.

Jyoti has received numerous awards and recognition over the course of her career, including the top 50 Asian Americans in Business Award from the Asian American Business Development Center, the top 100 Responsible Leaders from New York City and State, the Inspirational Achiever Award from Roshni Media Group, the Women's Leadership Award from the Council of Urban Professionals, and the Tribute to Women Honoree Award from the YWCA Princeton.

Chrissa Pagitsas: Jyoti, your career spans many industries, from non-governmental organizations to banking to consulting to hospitality. Tell me about your professional journey and what drives you.

Jyoti Chopra: As a longtime diversity practitioner, my work began in the public sector at the United Nations. In the late nineties, I spent a number of years at UNICEF, also known as the United Nations Children's Fund, and had a firsthand view of the inequalities that existed around the world with a special lens on gender issues. I was at UNICEF during the Rwandan civil war and the Bosnian conflict, where women and girls were used as weapons of war. I had a visceral firsthand view of issues in and around development, inequality, and discrimination. In my own home country of India, I looked at issues such as female infanticide and equal access to education for girls. These experiences had a very important part in shaping my outlook and spurring my interest in addressing diversity issues. It's something that was very deep-rooted and has stayed with me ever since.

Early on in your career, when you're influenced and shaped by defining moments in history, it leaves an indelible mark on you. In my case, it was this resolve that no matter where I work, what I do, that I want to have a meaningful impact. I want to be able to transform the lives and livelihoods of people in some way. This notion of impact got sealed in my career early on.

When I pivoted from the public sector into the private sector, I went from the world's most diverse organization, the United Nations, where I worked with people from over 175 countries or nations at that time and everything was concurrently translated into six languages, to the heart of Wall Street. I joined Merrill Lynch, one of the top three investment banks in the world, and went into the heart of retail wealth management, which was the polar opposite of the United Nations. It was not diverse. It was heavily white male-oriented, particularly at senior levels. It was a hard-charging sales, entrepreneurial culture, and environment, but it proved to be a great learning environment where I could pilot, test, and incubate a lot of ideas, products, and services for the company and its client base.

My journey in the private sector began under the auspices of James Gorman, who today is chairman and CEO of Morgan Stanley. He ran the wealth management business at Merrill Lynch at the time and was instrumental in re-engineering the business, segmenting the client base, and redefining the operating model. He is an incredibly progressive leader and challenged our team to examine new business opportunities around multicultural markets. The notion was to research and study diverse communities, develop an understanding of their needs from a financial and investment perspective, and develop appropriate and relevant service offerings. James seeded our new multicultural business development group, and we launched the business in 2001. In just a few years, we were able to build a highly successful and profitable business and one that became a model for the industry.

After a decade there, I went to work at Deloitte for four years as the global leader for communications and external relations, supporting Deloitte's global CEO and executive team. From there, I was recruited in 2012 by Bank of New York Mellon as the chief diversity officer to stand up a global diversity and inclusion strategy and build the function. Two years into that journey, there was an organic corporate restructuring of the organization, and I was asked to take on sustainability and philanthropy in addition to my existing scope of diversity and inclusion. That's really where my portfolio expanded to include sustainability and environmental, social, and governance work. It was an important turning point in my career because I had to look at things from an impact perspective from both the human and the social dimensions. I like to think of it in terms of the deployment of human and social capital.

In 2019, I joined MGM Resorts in Las Vegas for an incredibly exciting opportunity to lead people and social impact initiatives. I was drawn to MGM Resorts by the power of its brand, its incredible focus on guest service, and the sheer complexity and magnitude of the scope and scale of the company's operations and businesses spanning gaming, hospitality, and entertainment.

Pagitsas: What is the scope of your role at MGM Resorts?

Chopra: At MGM Resorts today, I have four areas under my responsibility— human resources, philanthropy, diversity, equity, and inclusion, and sustainability. Our philosophy centers on embracing humanity and protecting the planet. The combination brings to the forefront this notion of harnessing the power of both human and social capital. At MGM Resorts, we've anchored our work in this combined area around three main pillars. The pillars of our framework are fostering diversity and inclusion, investing in our communities, and protecting our planet. We have fourteen publicly declared long-term goals mapped to the pillars that run through 2025, along with two newly approved climate change goals that run through 2030. Examples of our 2025 goals include spending at least ten percent of biddable domestic procurement with diverse suppliers, donating five million meals through our Feeding Forward program, and reducing carbon emissions per square foot by forty-five percent and water per square foot by thirty percent from a 2007 baseline. Our 2030 goals are to halve our absolute global scope 1 and scope 2 carbon footprint and to source 100 percent of our renewable electricity in the United States.

Pagitsas: What governance structure has been built to oversee this framework and the sixteen goals?

Chopra: Ultimately, this work and accountability for it roll up to the Corporate Social Responsibility and Sustainability Committee of the MGM Resorts' board of directors. The CSR and Sustainability Committee of the MGM Resorts board has been in place for many years and has had a very specific focus on this domain area. I report directly to our CEO and president, Bill Hornbuckle, so there's a direct line of responsibility from the board to the CEO to the chief people, inclusion, and sustainability officer. Bill is a major proponent and advocate for our work in this area and is deeply engaged in all aspects. Additionally, I serve as the corporate liaison to the chair of the board committee, Rose McKinney-James. Furthermore, against our long-range goals, we have designated executives who serve as sponsors and champions for each of the goals, and we measure progress using specific key performance indicators.

Pagitsas: My takeaway is that you have clear lines of accountability and ownership of the sustainability strategy from the board down and across the company. Why does this structure work for MGM Resorts?

Chopra: It gets to this notion of what we refer to as shared ownership. We operate very much as a collective as we are indeed a collection of multiple businesses within multiple properties in Las Vegas and at regional locations. We are in the gaming business, the hospitality business, food and beverage, and venues. We have entertainment and retail businesses that span live and digital. We have a dozen resorts, the T Mobile Arena on the Las Vegas Strip, and multiple properties in different regional geographic locations. In order to effect a lasting, meaningful impact against goals, you must have a distributed model and a structure like ours. Today we are over 59,000 employees in twenty-nine locations around the world.

There is no way that a small team at the center of an enterprise of our scope, scale, and magnitude can effect the work that we're talking about from the center. You have to examine how you get buy-in, adoption, and acceptance in a distributed model so that you can drive meaningful activities and actions across the enterprise. In consultation with our CEO and our executive committee, we designated executive sponsors based on their areas of domain oversight. For example, our chief operating officer, who oversees all properties and our centers of excellence, including our food and beverage, is the executive sponsor for the goals over which he will have the most influence. Similarly, our SVP of human resources is involved in overseeing and championing the goal that relates directly to human capital and people. So, in these cases, the executives respectively serve as lead or co-executive sponsors to support our goals to exceed $120 million in cumulative employee donations to the MGM Foundation and to achieve seventy-five percent participation in domestic employee donations to the MGM Foundation.

We picked leaders whose areas of oversight and domain correlated with the programmatic pieces that would drive the success of their respective goals. That's how we developed our model.

Pagitsas: How do you engage with Rose and Bill on a day-to-day basis? And how do you partner with them to connect MGM Resort's impact strategy to employees?

Chopra: First, having a reporting line to the head of a board committee who is a subject matter expert in this area, has spent a lot of time throughout her distinguished career in the sector, and is very steeped in the domain, is incredibly helpful and significant. Rose and I collaborate directly monthly, and we set the agenda for the CSR committee together with input from MGM Resorts' CEO. We preview all the work products and materials with her before it goes into the committee session. Our CEO attends every meeting of this committee. So, in this structure, there is a direct line from the chair of the committee to the CSO and the CEO. I think it is significant and has propelled engagement from our management team and our property presidents.

The second strategic approach has been the involvement of our committee chair in broader strands of our work, like diversity, equity, and inclusion. An example is our courageous conversations forum. Bill and I launched it last year to curate heart-to-heart safe spaces for our employees to meet with management and to share what was on their minds and how they were feeling about the external environment and social justice issues.

The forums started in the aftermath of George Floyd's murder. We began with a courageous conversation cohort of African American colleagues. Rose joined us for that inaugural session, and she's attended every forum held since. To have a board director and your CSR and Sustainability committee chair actively engaged in leading the discussion and dialogue with the CEO and employees of the company, it's very powerful and quite unprecedented.

This allows your board member to have a direct line of sight and the ability to hear directly how employees are feeling. It was very powerful. What we ended up doing was extending that conversation to other cohorts. We now have a courageous conversation Asian cohort, a female cohort, and a Latinx-Hispanic cohort. The sessions range in size from thirty to fifty participants and include employees at all different levels, from the front line to the back of the house, with a range of tenure and experiences representing the regionals, Las Vegas, and corporate functions. They're very diverse in composition and purposely so.

In addition, we've created working groups from each of the cohorts. They developed action plans and defined the three to five areas that we want to tackle as a cohort to help us create a more inclusive environment and culture in our company. That has been very powerful. Bill is an active champion of our work and holds our leaders accountable in a variety of ways.

Chopra: The other thing that's been remarkable about Rose and Bill is they've not been afraid to issue public statements. We've done a lot of public advocacy work for LGBTQA+ rights and for the Black Lives Matter movement as examples. Our board and CEO have issued or co-signed statements on topics that we have felt important for us to take a position on. That's a great example of how to engage with your board on advocacy work supporting social impact and sustainability strategy and goals.

Pagitsas: With such a strong governance structure, committed board and senior leadership, and a shared ownership model, it seems like you can execute on planned sustainability and impact goals. Does the model also help you still create a positive impact when unexpected events occur in the market?

Chopra: One of the most recent and compelling examples is what happened around our food inventory during the onset of the COVID-19 pandemic. When the pandemic took off in early 2020, we made the decision to close all our properties on the Las Vegas Strip, throughout the United States, and in Macau. We literally made this decision in a matter of days. We had massive quantities of food and produce on property at every location. This is food that had been brought in for restaurants, banquets, events, large-scale conferences, and hotel bookings.

We had to determine very quickly how to get the food out to those who needed it most because there was a maximum capacity that we could store or freeze. With fresh food that would perish, we had a very tight timeline, a matter of days. Through our COO, we worked directly with our property leadership, chefs, and heads of food and beverage to figure out what food was going to perish, calculate our maximum storage or freezer capacity, and identify which produce and how much needed to be put onto pallets and distributed or transported out of our hotels and into community centers or food banks.

Then we had to find food banks or food charities that could take the enormous quantities of produce that we had and distribute it to communities in need. In Las Vegas, for example, we worked with Three Square, which is one of the local food non-profit partners we work with. We were able to get to them thousands of pounds of food to distribute out to homeless shelters and families in need within the community. All of this had to be mobilized in a very short period of time. Support from our leaders, from the CEO to the COO to the president of a property to the head of food and beverage to the chef who oversaw the kitchens, was critical to the success of the distribution.

In the middle of April of 2020, we organized a hot meal food distribution program through Catholic Charities. Our chefs and staff from the ARIA and Bellagio hotels came into the kitchen every single day for more than two weeks to cook hot meals, package the meals, and load them onto trucks which were then distributed by Catholic Charities to the community. Remember, this is at the early onset of the pandemic. Our kitchen staff was masked up, gloved up, stationed safely spread out, and packing up food. We were preparing over 1,000 meals a day in that period from the middle of April through the end of April, or almost 14,000 meals in total. This is an example of mobilization and thinking strategically about your resources to be most effectively used in a period that was, obviously, very difficult and very traumatic.

Thanks to the tenacity of our chefs and our long-standing community partnerships, MGM Resorts donated more than 662,000 pounds of food—or 552,000 meals—to the communities in which we operate. In Southern Nevada, we donated 444,000 pounds of food, equivalent to 370,000 meals. We donated an additional 219,000 pounds of food, equivalent to 182,000 meals, throughout the rest of the U.S.

When I think about the intersection of sustainability and the context of MGM Resorts and our hospitality industry, its greatest positive impact happens when one looks at resources and capabilities, identifies areas of greatest need, minimizes energy and waste, and puts them to their best use. Our impact was leveraging our core resources such as properties, food, and beverages, investing in the community, and making sure we didn't waste food and instead shared it with others at a time of incredible need.

Pagitsas: You are hitting on a common theme for sustainability executives, which is that often for the work that we do, there is no roadmap. That ability to innovate and create, sometimes in the moment, is an important scope and skill set of sustainability and ESG leaders.

Chopra: Yes. In the example of the scale of the food distribution during COVID-19, there was no blueprint for what we were doing. We had the infrastructure in place in terms of communication channels and the ability to quickly inventory our food and calculate how many pallets of food we were going to have. Those pieces were in place, but some of them were developed in the here and now. That's an important message in this story. You don't necessarily have to have large blocks of infrastructure in order to effect meaningful, sustainable impact.

Yes, it helps to have goals, and yes, it's important to have a governance model, but in this situation, we figured everything out in a matter of a few days and hours. What was the problem we were trying to solve? Getting perishable food out of numerous properties as quickly as possible before it went bad. What did we need? We needed transportation. We needed pallets. We needed a place for it to go. Some of it was done in the moment.

I'd say it's a hybrid situation where we used existing infrastructure and innovated and adapted quickly. I think it's a misconception that you need lots of infrastructure behind you in order to effect or implement meaningful sustainability initiatives. Some projects, like our Mega Solar Array, yes, are multi-year projects with large long-term investments. The solar array is a major project we pursued with our partners Invenergy and AEP, both of whom are large energy firms. There are some things like that, but then there are other things where you can effect change, and you can have an impact if you mobilize your resources and harness your supply chains in smart ways.

Pagitsas: What are the other critical leadership skills or philosophies a chief sustainability officer must possess?

Chopra: When you come into a role like this, history matters. As a sustainability leader, you need to understand the historical context of the journey around sustainability and sustainable practices at your organization. You need to do an assessment and an evaluation. I'm a big advocate for materiality assessments as a tool to identify where you are and how you should prioritize your work in and across the different areas of sustainability. It's important to understand and define who are your most important stakeholders, inside the organization and, if relevant, outside the organization.

Next, stakeholder engagement has to be a part of your model. That's a critical piece of successful sustainability leadership. It's understanding and defining your stakeholder group, engaging with them, and understanding the levers. It's a big piece of work from a leadership perspective. Standing up the governance model, as I described, is a critical component of stakeholder engagement and functional maturity. Those are the pieces that anyone coming into a role or standing up a function should consider.

Once you start there, you can look at your organization's role in the ecosystem of areas that are important. If I use MGM Resorts as an example, our ecosystem relates to our businesses, our operations, and how we engage with the world writ large. Therefore, our sustainability scope addresses things that matter to operating thousands of square feet of hotel, conference, and entertainment space. It includes solar energy, waste materials diversion, heating, ventilation, air conditioning efficiencies, water conservation, and LED [light-emitting diode] lighting, as examples.

Pagitsas: What are other examples of how MGM Resorts incorporates sustainability into its ecosystem?

Chopra: Since we host events attended by thousands of people annually, we have a major program around sustainable events and a team that works with our convention clients to deliver sustainable conferences and conventions. For example, when Amazon Web Services comes to one of our properties to host their annual conference, they ask us to work with them to make sure that they are delivering a sustainable conference. A little over a year ago, we worked with them to develop fully compostable lunch boxes. All of their luncheon meals for the duration of the Amazon Web Services conference were fully biodegradable and compostable. That's a big part of our portfolio.

Other examples are within housekeeping, our diverse and environmentally friendly supplier purchasing, and materials waste and food waste initiatives. With housekeeping, we now have an option for guests not to have sheets changed on a daily basis to conserve water and to protect the environment. These areas link to our supply chain, our operations, our business, and our role in the ecosystem. We've chosen these areas because they are material issues to our businesses, as identified by the materiality assessment. There is also a correlation between the products and services that we offer, sell, or deliver into the marketplace and the experiences we want our guests to have.

Pagitsas: What are critiques that you and your team field about the sustainability strategy, goals, or practices at MGM Resorts?

Chopra: Some of the criticism we get, interestingly, is around recycling, recyclables, and food waste which are visible and tangible to guests or customers coming in. For example, we don't have recycling bins in our offices, dining halls, hallways, or restaurants, but recycling very much takes place. It's all done in the back of the house. Bottles are separated from plastics and food waste, and even food waste is separated from liquids. Liquids are separated into fuels that can be recycled. There is a whole engine that happens in the back of the house at our properties that guests and customers often don't see, but we do field letters asking why we don't have recycling bins. We've got signs in the hotel rooms about laundry options and bedsheets, linens, and towels. That's one of the criticisms that we receive, but there's a lot of work that is being done behind the scenes, and it is operationalized.

Another area for which we receive criticism is gaming. We are in the gaming industry. We operate and run casinos in many jurisdictions and locations and are in full compliance with gaming regulations and laws. However, we're conscious of the fact that gaming can be a pathway to different forms of addiction. We have a big focus and commitment to what we term responsible gaming. We donate, invest, and provide training, resources, and education around game sensibility and responsible gaming. Additionally, we have partnerships in and around the area of responsible gaming. We have strategic alliances, partnerships, and contributions with many organizations, including the International Council on Problem Gaming, the American Gaming Association, Global Gaming Women Continuing Education Scholarship, and the Nevada Council on Problem Gaming. It's our intention to be responsible stewards for the industry and to run our businesses in a very responsible manner that safeguards the well-being of our customers, our guests, and the future generation.

Pagitsas: What is the property footprint of MGM Resorts? Is there a regionalized cultural approach to the sustainability and diversity strategy that you've designed to address the customers and operations in each region?

Chopra: Our footprint encompasses multiple locations across the United States includes a presence in Macau with two properties. We have a strong partnership with MGM China which independently runs properties in China. While we and MGM China publish independent sustainability reports, respectively, we have very close collaboration and transparency in terms of sharing information about how our reports are being developed. We've integrated Macau into some of our reporting to have global reporting. That's one dimension of how we're taking a global, joint approach to coordination, communication, more integrated reporting, and collection of data.

My view is that in the space of sustainability and in areas like diversity, equity, and inclusion, you need an overarching global umbrella strategy and approach for your organization, where you frame out the vision, the goals, KPIs [key performance indicators], strategic pillars, and what it is you intend to stand for, and you design and build your framework in such a way that you allow for regional and local adaptation and customization. Whether that's around diversity, climate, buildings, or corporate development, you allow for that localization to happen in market, in country. That's the essence of our approach. There are certain things that we might do in Vegas that are driven by the climate and the environment that we're in, whether that's around energy, electricity, water conservation, but that might differ for a property like, let's say, National Harbor in Maryland. It's a very specific design for the environment. In my view and drawing on experiences derived from leading international companies, having global, regional, national, and local approaches allows for the best-integrated diversity and inclusion strategies and engagement.

Pagitsas: It sounds like each property has responsibility for its share of the sustainability strategy, as appropriate for their geography. What projects are being implemented that are specific to the Las Vegas market, which has the largest footprint in your portfolio?

Chopra: Our Mega Solar Array is a great recent example. In 2018, we announced that we were commissioning a 100MW solar array in partnership with Invenergy North America. The idea was to develop and build a solar array in the desert as part of our long-term climate strategy. We opened the MGM Resorts Mega Solar Array in June of this year, 2021. It's a landmark and a milestone for us on a number of fronts. It's located in the desert in an area in North Las Vegas. It features approximately 323,000 solar panels that are arranged across 640 acres, which power and produce about ninety percent of the electricity for our daytime power needs on the Las Vegas Strip. That's a lot of green electricity for the over sixty-five million square feet of buildings that MGM owns on the Strip, representing thirteen properties. Just to give you another sense of scale, our Las Vegas portfolio has about 36,000 hotel rooms. The array's electricity production is equivalent to the energy used by about 27,000 homes in the United States.

We are the sole user of the energy generated by the solar array. It's a big stake in the ground for us and a big commitment. The array will help us accelerate progress toward our 2025 goal to reduce the company's emissions by forty-five percent per square foot. And, when we officially opened the array, we announced two new long-term goals—to halve our absolute global scope 1 and scope 2 carbon footprint and source 100 percent of our renewable electricity in the United States by 2030. These are big ambitious goals that represent a real commitment to climate action.

Pagitsas: Clearly, there is a carbon emissions and energy use reduction benefit. Does MGM also receive a financial benefit?

Chopra: There are definitely benefits in terms of hedging our exposure to energy price fluctuations. But most importantly, what drove the decision was our long-term commitment to reducing our carbon footprint and supporting renewable energy in a meaningful way.

Pagitsas: Looking ahead, what conversation are you having with the board and the CEO about ESG changes MGM Resorts needs to prepare for?

Chopra: First and foremost, the dominant conversation is our strategy around climate change and risk mitigation. We need to understand our exposure to climate change. This is an area many companies are grappling with the best way to go about it. How does one think about it? The obvious construct is geography. I'll use Beau Rivage, our property in Mississippi, which is located right on the Gulf Coast, as an example. In the relatively recent past, when Hurricane Ida came through, we had to close it down due to significant water damage. It's also been impacted by prior storms and hurricanes.

However, there are other climate-related risks that we may not have thought about yet or haven't experienced. I'll use Hurricane Ida as an example again. We have two properties in the northeastern region of the United States—one in western Massachusetts and one outside of New York City. I don't think any of us were expecting the level of flash flooding, the disruption, and all the debris that occurred as the remnants of Ida went through the Northeast region. Even more devastatingly, we weren't expecting the tragic loss of life that occurred in multiple states as a result of the storm. While flooding occurs on the Gulf Coast in Mississippi as a consequence of a hurricane, to have extensions of the same hurricane reach, impact and cause severe losses or business closures several states away is now something we have to plan for and manage. We're thinking about how to tackle climate change, climate strategy, and the associated risks.

Another area that is a big focus for us is the growing demand by institutional investors and shareholders in and around accountability for sustainability and sustainable practices and the coalescing by investors around the right frameworks to be paying attention to. We conducted a gap analysis of twelve different ESG frameworks, including Institutional Shareholder Services [ISS], Sustainability Accounting Standards Board [SASB], Dow Jones Sustainability Index [DSJI], Global Reporting Initiative [GRI], Task Force on Climate-Related Financial Disclosures [TCFD], MSCI, Sustainalytics, CDP, and others to assess our capabilities and current reporting. We thought about how we wanted to approach the analysis. How do we want to think about the results? It looks like the space is coalescing around SASB and TCFD, but even both of those are very different from each other in their respective methodologies, metrics, and KPIs.

As I look ahead, one of the aspects that I'm wrestling with is, how are we thinking about frameworks? How are we prioritizing them? Which ones do we want to pay attention to, and which ones will matter five years from now?

Pagitsas: These are important questions to tackle because a lot of time is needed to answer twelve different frameworks.

What are other themes beyond ESG reporting frameworks are investors bringing up?

Chopra: I partner closely with our investor relations team on investor roadshows and sessions. Broadly, what I'm hearing are the following themes. First, do you have a framework? Do you have goals? Do you have KPIs and practices in place to measure, track, and report? Second, do you have a robust enough governance model infrastructure in place? Third, where does accountability for this work ultimately reside and sit? Diversity and inclusion come up a lot. How are we doing on diversity, equity, and inclusion? Do we conduct gender pay equity analysis? Those are some typical areas institutional investors are looking at, in addition to board composition, board succession plans, and accountability at the board level for diversity, equity, and inclusion and sustainability work. I'd say those are the dominant prevailing themes from our conversations.

Pagitsas: I can see why investors are asking about diversity and inclusion because people are a critical resource for the hospitality industry. Without people, there wouldn't be entertainment or hotel operations. Therefore, what is a key change related to people or social issue you are anticipating in the hospitality industry?

Chopra: The most important social issues that we're grappling with center on changing ways of working, identifying hiring patterns, building a more agile workforce, and dealing with future hiring needs. That's turned out to be a real issue and challenge throughout the pandemic. People are just making different choices. The headlines around the Great Resignation are very real. People are making different lifestyle choices. Therefore, the employer value proposition becomes critically important, and the ways of working, benefits, and all of that come into play.

From a social perspective, what I'm contemplating for the next few years is how do we retain our workforce and not lose talented people? How do we hire and create jobs in a more agile manner? For example, we may have a shortage of lifeguards for the next two months and then have a shortage of guest room attendants or front desk agents. Is there a way to create models where people are trained and can serve in multiple capacities? Thinking about new work modalities, including ways of recruiting and ways of retaining people, is very much top of mind.

We're also grappling with remote, hybrid, and on-property work. We've had to change a lot in the last year. We were an organization that was pretty much 100 percent at work on property or in the office. We now have many employees that are 100 percent remote. We have some that are hybrid. But this is the hospitality industry, and our ways of working have dramatically changed.

Coming to terms with the office of the future, the hotel of the future, and the property of the future, envisioning and understanding it is critical. What could we potentially automate or leverage artificial intelligence and robotics around? If we digitize a menu and people order from their iPhones, it means we don't have to print paper menus anymore, and those resources can be redeployed. It's worth thinking about a lot of things like that. How do we truly harness the power of people and have them focus on the experience, the guest interactions, and the curation of meaningful experiences so that people want to come back and revisit our properties? That's what we're thinking and spending a lot of time on.

Pagitsas: Lastly, do you view these as fairly seismic shifts in the hospitality industry? What are you most looking forward to with these shifts?

Chopra: I believe it's an exciting time for us. If you just think about check-in historically, in the past, everybody had to go to a front desk to check in, or maybe you had a VIP check-in service. Well, now we have mobile check-in where you don't have to stop at a desk and wait for a person to assist you. You can take your phone, check in, and have keyless contact access to your room. Think about that one simple example and how it's changed the guest experience.

The guest can now access their room faster and more efficiently. You can redeploy your human capital to other areas that are more meaningful and focus on different aspects of the guest's visit and experiences at your property. There's exciting potential in rethinking the future of hospitality and the future of gaming and what that could look like in the years ahead.

Emma Stewart, PhD

Sustainability Officer
Netflix

Emma Stewart, PhD, is Netflix's sustainability officer. Netflix is a streaming entertainment service with 222 million paid memberships in more than 190 countries with $30 billion in revenue as of 2021. Emma leads the enterprise's strategy to be net zero by 2022, raise awareness of environmental sustainability through film and television content, and engage millions of members on climate and environmental change.

Emma brings twenty years of expertise at the intersection of environmental and business strategy. Among her prior leadership roles in sustainability, she founded the first Sustainability Solutions product group at Autodesk, led the Urban Climate & Energy department at World Resources Institute, and established and ran the research and development division at BSR, which serves more than 300 Fortune 1000 companies.

Emma was an architect of the first tools for setting science-based climate targets, which more than 1,000 global companies have adopted. She created and taught the

© Chrissa Pagitsas 2022
C. Pagitsas, *Chief Sustainability Officers At Work*,
https://doi.org/10.1007/978-1-4842-7866-6_7

Intrapreneurship for Sustainability courses at the University of California, Berkeley, and Stanford University business schools.

Her work has been cited in international publications, including The Economist, The Wall Street Journal, *and the* Harvard Business Review. *She has been named "one of the most powerful women under 45" and an "urban pioneer" by* Fortune *magazine and a "sustainability insurgent" by the* MIT Sloan Management Review.

Chrissa Pagitsas: Many readers of your chapter are likely subscribers of Netflix and are very familiar with Netflix's shows. However, only a handful may be familiar with Netflix's sustainability strategy. Would you share with us Netflix's overall sustainability strategy?

Emma Stewart: In March 2021, we announced that Netflix would achieve net-zero greenhouse gas emissions by the end of 2022 and every year thereafter. Our Net Zero + Nature strategy has three primary components. First, we will reduce our internal emissions and address indirect emissions in partnership with our suppliers. Second, we will retain nature's existing carbon storage, protecting ecosystems like forests that prevent carbon from entering the atmosphere, and invest in other high-impact projects aligned with climate science to address our indirect scope 3 emissions. Third, we will remove carbon from the atmosphere by restoring and regenerating natural ecosystems. We'll focus especially on projects that advance sustainable livelihoods, biodiversity, climate resilience, and environmental justice.

To support this strategy, we joined the United Nations Business Ambition for 1.5°C group of companies, as well as America Is All In, a consortium committed to execute against the Paris Agreement's goal to limit global warming to 1.5 degrees Celsius. It's also critical that we connect with our customers to partner with businesses to advance action toward climate issues.

For our customers, we've curated more than thirty titles that cover the real-life beauty of Earth's life-support system into our Together for Our Planet collection. Shows, like *Our Planet* and *Breaking Boundaries,* cover the real-life beauty of Earth's life support systems and the science of planetary boundaries while spellbinding thrillers like *Ragnarok, Green Frontier,* and *IO* captivate viewers' imagination. Titles like *Down to Earth with Zac Efron* and *The Minimalists* show us hopeful ways to advance healthy and sustainable lifestyles while *Penguin Town* and *Izzy's Koala World* delight the entire family.

To connect and partner with businesses, we are a founding member of the Sustainable Aviation Buyers Alliance, which works to scale up the supply of sustainable aviation fuels, and the Business Alliance to Scale Climate Solutions, where we team up with companies like Google, Amazon, and Salesforce on investments in nature-based solutions and carbon removals. These are just some of the initiatives that combined will drive forward the positive environmental impact we seek to have on the planet.

Pagitsas: Let's touch on Netflix's nature-based solutions before diving into the net-zero goal. Why are they critical for our planet's future?

Stewart: Without taking these steps, it's as if you were pouring wine into a glass, but the glass had a hole in the bottom. That's effectively what society is doing today by letting natural carbon sinks die off or burn down. Just this year the Amazon rainforest turned from carbon sink to carbon source due to rampant deforestation. We are burning forests and letting that carbon escape up in the atmosphere. So, using the wine metaphor, we're essentially going to be thirsty and out of money.

Our approach is to protect nature's well-proven capacity for retaining and removing carbon from the atmosphere, which is ultimately what allowed humanity to settle, become agriculturalists, and industrialize. The stability that we've seen over the past ten thousand years in the climate is largely due to natural systems.

Nature is at the heart of our commitment. As environmental leaders like Christiana Figueres tell us, we can't achieve climate goals without protecting and regenerating natural ecosystems. This approach buys us time to decarbonize our economy while restoring these life-support systems. For example, the Lightning Creek Ranch project in Oregon shows our "retain" goal in practice. Our investment there is helping preserve North America's largest bunchgrass prairie. In Kenya, we're supporting the Kasigau Corridor REDD+ Project, protecting the dryland forest that's home to hundreds of endangered species, and providing local residents alternative incomes to unsustainable activities like poaching.

Pagitsas: Let's touch on the net-zero goal and the 2030 goal. What is the ambition and scope of these goals?

Stewart: By the close of 2022, Netflix will achieve net-zero greenhouse gas emissions. By 2030, we will reduce our scope 1 and 2 emissions by forty-five percent below 2019 levels. To have the greatest positive impact in our industry, we will voluntarily take responsibility for more of our supply chain or scope 3 emissions than required by the Greenhouse Gas [GHG] Protocol.

Unlike most climate targets in the market, we decided to include most of scope 3, which are indirect emissions from the supply chain, not just business travel and commuting, as most companies do. Netflix has voluntarily taken responsibility for the scope 3 categories from our largest sources, including purchased goods and services, Netflix-branded licensed and partner-managed production energy, to our smaller sources, including corporate commuting, corporate capital goods, production travel, Open Connect network, and Amazon Web Services.

Pagitsas: According to your analysis, roughly half of Netflix's carbon footprint lies within the physical production of the stories we watch. How does the net zero goal get integrated within the production process of Netflix's shows and movies?

Stewart: My team likes to joke that I don't have a tattoo, but I should get one because I am such a broken record about how the order of events matters deeply when it comes to achieving this target. "Optimize, electrify, decarbonize" is the most economical and environmentally beneficial order of events. If you skip to the end, you're going to cost yourself a lot of wasted money and effort.

The first order of events is to optimize energy use. All too often, people leap right over energy efficiency because it's invisible, even though it has a stronger business case than the other two steps. So, for production, what does that look like? That looks like using LED [light-emitting diode] lighting, which itself is seventy-five percent more energy-efficient, but also doesn't create a bunch of waste heat which we then have to cool for. Lo and behold, many actors like it because it doesn't melt their makeup like conventional bulbs used to.

Pagitsas: That's a classic sustainability "win-win" scenario where an environmental benefit has a business benefit.

Stewart: Yes, that's right. It also turns out that a lot of studios don't have sufficient power from their local utility to power their stages or sets. They end up compensating for that with auxiliary power that they bring in. What's readily available from the vendors right now is diesel generators.

So, step one is to make sure you have energy-efficient measures, such as the LED lighting, heating and cooling systems, and other appliances that use a lot of energy. But energy efficiency extends to thinking about our transportation and catering too. We use local crews wherever possible, which cuts down on energy-using travel—and creates local jobs—and donate leftover catering food. Food waste produces methane. Step two is to make sure that that energy load is as fully served by the local utility as possible by predicting the full energy load, including peaks, and ensuring the utility provides sufficient power.

After these two optimization steps are completed, the next order of events is electrification. Electricity is a cheaper fuel per unit of power than liquid fuels such as gasoline or diesel. Electricity is also much easier to green than liquid fuels. Our priority is swapping out diesel generators with mobile batteries used to power the set and using electric vehicles to transport actors and the filming crew to and from the set or the location. To achieve this goal, we need to upgrade and enhance the power coming into that set, which, as we discussed earlier, is often lacking. Then electrify all equipment being used at the facility, if possible. We're experimenting with both on multiple productions and plan to partner with vendors to scale up supply.

The last order of events is to decarbonize whatever remains of the fuel and electricity that you need. Let's say you're reliant on diesel. Well, let's get some renewable diesel in there. Let's say you need auxiliary power. Well, then let's make sure that it's a battery, a mobile battery, or potentially even green hydrogen in the form of a fuel cell. Our solution is to purchase 100 percent renewable electricity for all our operations where possible, even if it means paying a small premium to the utilities serving our California headquarters and London studio. Elsewhere, we're buying renewable energy credits to cover every location but one. We're also exploring direct investments in renewable energy projects.

Pagitsas: Next time I watch a Netflix show, I'll be thinking about all these energy-saving activities going on behind the scenes! Given the complexity of the net-zero target, how do you tell the Netflix sustainability strategy to your 222 million household subscribers globally?

Stewart: Netflix's net-zero's target design itself is rather complex and grounded in science, so we used a combination of technical blogs, an investor-ready environmental, social, and governance [ESG] report, and a short film, Netflix's natural storytelling medium, to explain the arc of our reasoning. This is arguably the most consequential decade in human history. Literally for millennia to come, our decisions this decade will be felt. So, how do you do that in an inspiring but also new and engaging way?

Pagitsas: Indeed, how do you do that?

Stewart: Our company's strength is in supporting creators to tell incredible stories. We are always looking for the right stories for our subscribers that entertain and educate on sustainability. We recently brought the story of climate change into millions of households with Adam McKay's new film, *Don't Look Up* [Netflix, 2021]. Starring Leonardo DiCaprio, Jennifer Lawrence, Meryl Streep, Jonah Hill, Cate Blanchett, Tyler Perry, Mark Rylance, Rob Morgan, and many more incredible actors, the film tells the story tells the story of two low-level astronomers, who must go on a giant media tour to warn mankind of an approaching comet that will destroy planet earth.

The comet hurtling toward Earth is an apt metaphor for climate change, but unlike comets, we have cool and proven ways to combat climate change. And every individual watching can be a part of the solution. In a recent piece for *The Conversation*, climate psychologists Dr. Barbara Hofer and Dr. Gale Sinatra said it best, "The most important difference between the film's premise and humanity's actual looming crisis is that while individuals may be powerless against a comet, everyone can act decisively to stop fueling climate change." ["'Don't Look Up': Hollywood's Primer on Climate Denial Illustrates 5 Myths That Fuel Rejection of Science," January 5, 2022.]

We know that this movie has catalyzed conversations around climate change in a way that few other reports and media have not to date. The movie has broken multiple Netflix records. While only released at the end of 2021, it's already the second most watched film in Netflix history and broke the record for most viewing in a 7-day period. While many people find climate change to be intimidating, we feel that we can make it relatable and actionable. Every time someone talks about climate change accurately and engagingly because of our stories, it means that we are, hopefully, inspiring someone to take action.

Pagitsas: Before shifting gears to Netflix's internal operations, I'd like to follow-up on the generous boundaries you've drawn for Netflix's net-zero commitment. Why did you decide to include emissions from productions that were licensed to but not directly produced by Netflix?

Stewart: Most people don't start with an understanding of where the boundary for climate targets should be. Even the standards from NGOs [non-governmental organizations] are unclear for media companies like ours. We did what we thought would be the most responsible thing to do and most logical for our consumers, which is, "If it's Netflix branded in their minds, it's Netflix's responsibility, regardless of whether it is self-managed, partner-managed, or licensed." Similarly, if we were only to focus on self-managed productions, we wouldn't have the volume of productions to get that supply chain to step up. If we cast a wide net and included those things that were being produced by others, everyone now has the shared problem of getting energy-efficient, cleaner technology into production.

Pagitsas: What is excluded from the net zero and science-based climate target boundaries and why?

Stewart: One thing that we don't include in the net-zero boundary are the emissions from the delivery of streaming entertainment to the consumer. That's because the GHG Protocol, used by all companies globally for carbon accounting, says that such emissions are considered the emissions of the Internet service provider or the emissions of the device manufacturer upon which you're watching Netflix. Regardless, we do want to understand what the true carbon footprint of streaming is. It has been a topic of great fascination, but there is also very poor data available on it. So together with a number of our peer companies, we've teamed up with the University of Bristol in the United Kingdom as well as academic advisors from Lawrence Berkeley National Laboratory, University of California, US Department of Energy, and the International Energy Agency to apply best practices in greenhouse gas accounting and life cycle assessment across multiple Internet-based products, like video streaming, advertising, and publishing.

It's all the same shared infrastructure. What we found was that streaming one hour of Netflix in 2020 would cost you, on average globally, well below 100 grams of carbon dioxide equivalent, which was far less than some of the estimates floating around. For people not conversant in that currency, that's the equivalent of microwaving four bags of popcorn or running a ceiling fan for four hours in North America. Our industry wants people to understand contextually where streaming sits in the relative set of activities that an individual might take on. The Carbon Trust has published a white paper on this topic, called the *Carbon impact of video streaming*, validating the methods and results. Now we can team up with companies in the streaming delivery supply chain, namely the data center companies, the Internet service providers, and the device manufacturers, to apply our collective muscle where it matters most, which is in the device use phase such as the electricity used by electronic devices.

That's something I was big on with my students at the University of California, Berkeley's Haas School of Business. I taught a class called Intrapreneurship for Sustainability: Driving Environmental Change from within Corporations. I told them, "Always apply the 80/20 rule." The rule generally states that twenty percent of your activities will result in eighty percent of your results. You need to focus on those activities that really matter. If you don't focus, you'll drive yourself crazy chasing these things that ultimately are immaterial. This work helps elucidate the true carbon impacts of streaming and the degree of operational control that a company like Netflix has over that.

Pagitsas: Looking internally, how are you addressing the footprint of Netflix in its corporate offices?

Stewart: With our landlords' permission, we've conducted extensive energy efficiency audits in the past few months and are now turning to implement their recommendations. In our corporate offices, I'm particularly excited about our new facility in Los Angeles called EPIC, where we worked with the landlord to cut energy use by fifteen percent above even the very strict energy requirements in California. It's also the first large office building in Los Angeles to use building-integrated photovoltaic panels!

Pagitsas: Shifting gears to organizational design, where does accountability for the sustainability strategy sit within the organization?

Stewart: The Sustainability Office sits under our CFO. The ownership of ESG reporting is held by our corporate controller VP, who also oversees our financial reporting. This helps ensure that our ESG reporting benefits from our financial and internal audit protocols. Additionally, we enlisted EY, our Independent Accountants, to conduct a limited assurance of our 2020 scope 1, 2 and 3 carbon emissions. Last but not least, we incorporated climate change risk in our enterprise risk management.

Sustainability and the sustainability officer's function are housed under the CFO for multiple business reasons. The finance team touches every part of the business, from operations, content, and product to marketing, communications, public policy, legal, and investor relations, enabling us to align business growth and sustainability. It is uniquely well-suited to early-stage planning and forecasting, critical to long-term success, and to quantitatively assessing the full range of benefits from ESG investments, from cost reduction, new revenue streams, and risk mitigation to customer retention. It ensures that GHG accounting is treated with the same processes and rigor as financial accounting. Last but not least, the finance team engages with our debt and equity investors directly and allows for integrated ESG and financial disclosure in the future.

Pagitsas: What are the benefits of this organizational design?

Stewart: Our structure and location within the organization enables us to make sure sustainability is part of smart and strategic business decisions across all areas of business. You have Netflix Studio, which is all the physical production and the creative storytelling elements. Then you have the Product division, which designs the platform experience for our subscribers and the backend systems that support it. Then you have the Content team, which acquires and develops TV and film, and Publicity and Marketing. There's almost no team that we don't interact with.

Pagitsas: You shared that the controller's office has taken on ESG reporting. What was the intent of that assignment?

Stewart: We intentionally asked the controller to take on ESG reporting because she oversees financial reporting. We want to reach that end state of having the same degree of rigor as financial reporting. Together, my team and the controller's team scan the market for important ESG events and changes— from pending changes from the SEC [US Securities and Exchange Commission] on disclosure to public comment periods on the Task Force on Climate-related Financial Disclosures [TCFD] and CDP.

Pagitsas: Looking at your background, you are unique among many chief sustainability officers in that you have a PhD in environmental science and policy and management. Why did you pursue a PhD?

Stewart: The PhD was intended to give me the luxury of time to "learn how to learn." I never intended to go into academia. I was in the minority, as you might imagine, with that viewpoint. Everyone in my cohort was planning to become a professor. But I didn't spend a lot of time worrying about how many publications I had.

Instead, I spent a lot of time concerning myself with which disciplines I needed to unite to advance sustainability in the corporate sector. When I started this academic journey back in 1996, I discovered that there was no degree program. Frankly, no faculty had looked at the intersection of the relevant disciplines. When I presented my dissertation research in 2004, I ended up designing my dissertation committee from four different schools at Stanford University— the business school, the civil engineering school, the Center for Environmental Policy, and anthropology. I had a civil engineer, a sociologist, a geneticist, and the editor of *Science*. I wove together the disciplines that we now know to be the sustainability field. That term—in the way we understand it today—didn't exist when I started the degree.

Pagitsas: What topic did you explore in your PhD dissertation?

Stewart: It was on corporate sustainability before that was a field. I saw this natural experiment ripe for the taking. It was about multinational companies operating in both the Dominican Republic and Cuba and largely in the tourism and hospitality industries, the primary economic drivers for both of those countries. In Cuba, you had extremely strict environmental laws and regulations and a very strong social safety net, along with lots of other political concerns. Yet, in the Dominican Republic, you had almost the opposite. It was a perfect natural experiment of juxtaposing the same companies operating in two island states in the same region of the world, attracting a lot of the same customers, and even having some of the same investors but within two different regulatory environments.

It turned out that a company's performance was highly dependent upon those regulatory contexts as measured by a set of eighty-seven environmental and social indicators I designed. There were some other surprising findings. For example, in Cuba in particular, you have a lot of very senior executives who have come from other fields. You see nuclear scientists running hotels, for example. I found a very statistically significant correlation of better environmental performance among the companies run by executives who came from non-tourism or hospitality fieldsfv. If they were a former mathematician or nuclear scientist, their companies performed higher than certain policies would require them to. If they had come through the traditional tourism and hospitality training paths, they tended to gravitate to the norm or below the mean. So those were a couple of things that I discovered along the way.

Pagitsas: What did your PhD experience teach you beyond the findings of your research?

Stewart: My dissertation involved a lot of fieldwork. I was out of the United States more than I was in the country. I had to teach myself Spanish. In Cuba, I was tracked down by the Cuban Ministry of the Interior. Guys in fatigues showed up at my doorstep in the middle of the night, asking me what I was doing there and how dare I interview these executives. There are lots of war stories from that period in my life. It taught me to be very resourceful, improvise well, and network within an inch of my life. All those things paid off in spades in the ensuing years.

Pagitsas: So, a PhD is not required to be a chief sustainability officer?

Stewart: No, a PhD is not a requirement. It happens to work out quite nicely for my position here at Netflix because our primary principle and what we hold ourselves accountable to within the sustainability office is to move fast but to ground everything in science. I am always bringing myself and my team back to that touchstone of, "What does the scientific literature already say?" If there's a gap in the literature, then we need to fill it by running a study or survey before we can move forward. Otherwise, we will do it wrong. We will expend huge cycles and money and effort on something that is not empirically proven, or we will find that the efficacy of our work is undermined. Therefore, science is where we always start.

Whether it's the behavioral science literature, the environmental science across multiple vectors, the economic elements of sustainability, the social and psychological science, especially as it relates to our consumer base, we steep ourselves in it. For example, there's plenty of empirical evidence in the psychological sciences about how people interpret climate information from the news and other sources, but almost no impact studies on film or TV sources. There are plenty of consumer-facing lists of pro-environmental behaviors, but few are grounded in the science of which actions have the most impact quantitatively. My team knows that I'm a broken record on that. In that sense, having a PhD works out because it would be a bit hard for me to continually beat that drum if I hadn't charted that path at least briefly. But, no, it's by no means a prerequisite.

Pagitsas: In addition to your unique achievement of a PhD in sustainability, your career spans working for large international companies to a think tank and running research and development for a non-profit. What are your professional takeaways from leading key strategies within a think tank and non-profit?

Stewart: I think cross-training is beneficial for any professional. For example, I'm far more effective in interacting with NGOs now because I've been in their shoes. I've been the one who has spent years raising the grant dollars from foundations to hire a team. I've been the one who has to answer to the board. So, I can empathize with their situation. However, I've also seen quite a bit of mismanagement, so I know what's realistic to expect in terms of their ability to work strategically or swiftly. Think tanks tend to gravitate toward intellectual people. They attract extremely bright analytical minds but who sometimes forget about the practicalities of getting something done on the ground.

Pagitsas: How would you contrast the think tank's approach to sustainability to a large for-profit company's approach?

Stewart: For the most part, the corporates for which I've either consulted or worked for internally don't typically think about engaging the academic or research world. Their first call is not to a professor at Yale, someone who I just got off the phone with twenty minutes ago. Their first call is to someone internally to ask, "What are our peers doing?" That's their measure of how they should design their strategy. Frankly, it's a little insular.

Happily, at Netflix, we are removing that wall between the research community and the corporate community. I would say two or three out of my ten meetings a day are with folks who dedicate their careers to deeply understanding and thinking empirically, whether they sit in an academic institution or a think tank or even in an NGO. We ask them to challenge my assumptions and to help us do things like design surveys. Or we ask them to push back on something in our strategy memo or run numbers accurately. That osmosis between the research community and the corporate community is something that I feel has been lacking and something that I'm so delighted that Netflix embraces.

This has manifested in things like our advisory group for sustainability at Netflix. It is composed mostly of people who have been in and around the scientific world, whether that science is atmospheric science or behavioral science, social science, and policy negotiators. It's the right and the left brain working together.

Pagitsas: Looking broadly across your corporate experiences, what key lessons have you drawn from them?

Stewart: One of the things I learned was to draw my executive influence map early on to find out who my key decision-makers were. You can't just look at an organizational chart and figure it out because it's not necessarily apparent. For example, when I joined Autodesk, one of my first calls was to the VP of government affairs. I said, "I just want to make sure that you're not lobbying against bills that my team would advocate for." He and I became close collaborators, and we walked the halls of Congress together.

One of my second calls was to the VP of real estate, in which I said, "Can we work together to green our own office real estate? That's ultimately what I hope Autodesk does with its software for customers." Autodesk creates software products and services for many industries, including those working in the design and construction of buildings, including the architecture and engineering industries. So, using the software on our company's real estate was a way for us to demonstrate the viability of new sustainability features within our software to our sales teams before bringing it to customers. However, at first he viewed my team as a pain because we kept asking for the utility bill and square footage information to calculate our carbon footprint.

Pagitsas: Why was your team considered a pain?

Stewart: We were a pain because we weren't providing him any value in those initial conversations. So, we stopped doing that entirely. Instead, I said to my team, "We need to show business value within two minutes of having a conversation with this person." My team and I then looked for opportunities to make him a hero because he viewed us as a cost center, and it was a time of cost-cutting at the company. So, we started using the data from the office buildings for which he was responsible to prototype ways to improve the Autodesk software. Together, our teams ultimately brought that new software to the market, and his team and purview expanded within the company.

This experience demonstrates how important it is to look for opportunities to put peer executives on the podium and have them do the talking because there's nothing like having to familiarize yourself with talking points to really digest them. I'll never forget this. The VP of real estate gave a talk that my sustainability team had set him up with, and he surprised me.

He said something to the effect of, "I'm sure most people, including Emma, assumed that my primary objective is to increase dollars per square foot, but I also care about decreasing carbon per square foot." I remember thinking, "Oh, that's a breakthrough." Those are the moments when you realize you have to draw the executive map to understand your peers' priorities. You also always have to be indispensable in these meetings before you ask for something because, otherwise, you're just a nag.

Pagitsas: It seems like you're also upending the conventional assumption that sustainability leaders are soft or not focused on business outcomes like profit or capital.

Stewart: Yes, that's right. Here's another example from my time at Autodesk. I was in a meeting with someone from HR [human resources] and my boss. We had wrapped up the meeting, and the HR person said, "It must be exhausting being the moral compass of the company." They were trying to be kind. I was kind of struck. I didn't know what to say. My boss stepped in and said, "Are you kidding? She's the hardest-nose business type in every conference room." I took that as a big compliment. If I were seen in this role delivering on the best interests of the company, particularly the financial best interests, then I would be taken seriously.

Today, we're in a different mode when sustainability is much more mainstream than when those conversations took place. But as the leader of sustainability, you still have to figure out the answer to the question, "Are these people just humoring me and the sustainability strategy because they have been told 'it's the right thing to do,' or do they truly believe that it can be a value generator for the company?" It is incumbent upon my team and me to continually be indispensable no matter what the needs of that executive might be.

In so doing, we get pulled into a lot more conversations. In contrast, some colleagues in the sustainability field knock on the proverbial conference room door and say, "I'm from sustainability. I should be in this meeting." Well, no, not if the sales, business, and marketing teams don't want you there. You have to make yourself so valuable that they can't hold the meeting without you.

Pagitsas: How do you show your business peers that you are valuable?

Stewart: I always say, "Always be the voice of the customer if you can be." My team at Autodesk got the reputation of being the team that was closest to the customer. That is a huge compliment when you think about it. Even the sales team referred to us that way which was remarkable because they're out with customers every day.

The first reason that we did that customer research was out of curiosity. We wanted to know what the customers needed. Second, if we spoke from the voice of the customer, it was unimpeachable. There was no debate. The response from the other teams would be, "Oh, okay, the customer needs these features. All right, let's go prototype those." However, that was a different era between 2008 and 2016 because now sustainably is considered mainstream and central to a lot of companies' well-being.

For a new CSO [chief sustainability officer] who has come from within sales or the core business, connecting with the customer and their needs probably comes naturally to them. It's not as obvious to someone coming into the organization from the outside with training in environmental science and who does not have a business background. They might think, "I probably just need to hammer on about sustainability and how important it is. This is dire. The world is burning." Those arguments tend to fall flat, or at least are self-limiting.

Pagitsas: It seems like you've identified a set of key skills for being a successful sustainability leader.

Stewart: Yes, I call them "the secret techniques of an intrapreneur for *K* sustainability." There is a set of logical techniques and a set of interpersonal techniques, and some of them overlap. Be indispensable, be occasionally exclusive. Co-invest even when you don't need to chip in. It unites you, and now you have a shared fate. Those are some of the interpersonal ones.

Another one is "channel your inner anthropologist." Anthropologists love talking about the power of reciprocity. It's essentially what built human society as we know it. The same holds true in a corporate organization. You should always bring something to the table and at least meet, if not exceed, what the other party brings. That might be in the form of something that you know better, like in the sustainability realm, or that you have brought in from another element of your professional career.

These were the fundamental "secrets" or lessons I taught my students in my intrapreneurship class at Berkeley. It's essentially an advanced course on the soft skills of creating change within a complex political organization like a corporation.

Pagitsas: What soft skills are you working on for yourself?

Stewart: One thing that I still struggle with and continually work to improve is letting my right brain tell stories. I tend to gravitate toward charts, facts, and figures. One of the things that Netflix is so good at is pushing me to tell more stories and make sustainability more relatable. I used to carry around a book called *Don't Be Such a Scientist: Talking Substance in an Age of Style* [by Randy Olson (Island Press, 2009)], which was my reminder that you cannot speak to everyone in those technical, scientific terms. You're going to lose most people. You're going to bore everybody. So, I ask myself, "Can you convert them into stories?"

Pagitsas: As new leaders enter the sustainability field, what is a relatively unknown skill critical to the success of a sustainability leader?

Stewart: The sheer technical depth of the field. People are often surprised by how quantitative, deeply analytical, and picky the sustainability field can be. For example, we were just talking to a communications agency that does nothing but corporate sustainability communications work, but within two minutes, we had lost them. The sustainability field is very jargon-prone. Yet, the jargon often has real meaning to it.

It's a highly multidisciplinary space, from engineering, economics, finance, and accounting, to biology, ecology, physics, sociology, and policy. I think people underestimate just how multi-faceted and then deeply technical it is. Because when you think about it, what is not relevant to the environment? Very few things. That means that you have to be comfortable talking about accounting because that's what carbon footprinting is. As the sustainability leader, I also need to know about SEC regulations, I need to know about macroeconomics when it comes to energy pricing, I need to know about utility regulations, all of these things, and that's just on the operational side. Then, how do you design a green building? Oh, and then how do you decarbonize production?

That's kind of the nitty-gritty. It's almost every discipline woven into this contrived field called *sustainability*, which is just jargon for "keeping the planet habitable." I think people don't have a full appreciation of just how remarkably complex that is. They assume that one person can represent everything.

Pagitsas: What are some of the most visible changes about sustainability's acceptance in the corporate world?

Stewart: I've got to say, I've never seen the momentum or the level of interest in the sustainability space. It seems to be snowballing. It's front-page news. You've got big investor firms plowing in capital and not just putting out talking points.

We had a job posting for a very junior position on the sustainability team that attracted over 1,400 applicants in five days. It's a very hot field. That's not to say it's immature or young. You have the old guard, the initial CSOs, many of whom I knew, who are now starting to retire. It's been a twenty-year lifetime for this field in the corporate world. I think the old guard would be proud now to see that the new guard is having a much easier time.

Pagitsas: On a final, personal note, what motivates you daily?

Stewart: From a personal standpoint, I used to joke with my team that I would take the subway into work and think, "We're doing nothing short of changing capitalism to be in harmony with the world's natural life-support systems." At the end of the day, I would come back and think, "We have eight years to keep the world to no more than 1.5 degrees Celsius. There's no way that's going to happen." I go through that mental roller coaster every day. But as the former CSO of Marks & Spencer reminded me recently, "Focus on what you can control." So, I'm grateful to have hired a world-class team, have handpicked the best partners in the market, and be challenged every day to bring both sides of my brain—the scientist and the storyteller—to the office.

Technology, Telecommunications, and Professional Services

Kara Hurst

Vice President, Worldwide Sustainability
Amazon

Kara Hurst is vice president of Worldwide Sustainability at Amazon. Amazon is guided by four principles: customer obsession rather than competitor focus, passion for invention, commitment to operational excellence, and long-term thinking. Amazon strives to be Earth's Most Customer-Centric Company, Earth's Best Employer, and Earth's Safest Place to Work. Customer reviews, 1-Click shopping, personalized recommendations, Prime, Fulfillment by Amazon, Amazon Web Services (AWS), Kindle Direct Publishing, Kindle, Career Choice, Fire tablets, Fire TV, Amazon Echo, Alexa, Just Walk Out technology, Amazon Studios, and The Climate Pledge were pioneered by Amazon. In 2020, the company had $386 billion in revenue. As of Q4 2021, the company had over 1.4 million employees globally.

Kara's responsibilities cover multiple environmental and social sectors, including renewable energy, energy efficiency, sustainable transportation, product sustainability, circular economy and waste reduction initiatives, social responsibility and responsible sourcing, sustainability science and innovation, and environmental compliance.

Before joining Amazon, Kara was the CEO of The Sustainability Consortium. Her previous experience includes working as vice president of BSR, where she led BSR's New York and Washington, DC, offices and the global partnership practice with governments, multilateral organizations, and foundations. Kara co-founded the

© Chrissa Pagitsas 2022
C. Pagitsas, *Chief Sustainability Officers At Work*,
https://doi.org/10.1007/978-1-4842-7866-6_8

Electronic Industry Citizenship Coalition, now known as the Responsible Business Alliance, and led as executive director of the public-private venture OpenVoice, building out early teen channel content for AOL and others.

In her early career, she held roles at the Children's Health Council, leading interdisciplinary education and development programs, at the Urban Institute as a research lead in the public finance and housing division, and worked in the offices of two elected officials—former San Francisco Mayor Willie Brown and the late Senator Daniel Patrick Moynihan (D-NY).

Chrissa Pagitsas: Kara, let's start with an overview of Amazon's sustainability strategy. What is it, and what are its primary goals?

Kara Hurst: Our strategy is driven by our desire to be the most sustainable option for our customers. Our sustainability strategy is centered on The Climate Pledge, which is our commitment to achieve net-zero carbon emissions across our businesses by 2040, ten years ahead of the Paris Agreement. Through this commitment, Amazon is investing in a range of large-scale solutions to mitigate greenhouse gas emissions in our business, some with immediate decarbonization impacts and others with longer-term payoffs.

It indicates the urgency with which we're addressing climate change because this is not just our own commitment but an invitation to any other company that wants to join us in partnership. Our hope is to get others to make that commitment along with us. As of Q4 2021, over 200 companies have already committed with us to The Climate Pledge, which is just phenomenal. Signatories include Procter & Gamble, Visa, PepsiCo, Verizon, Siemens, Microsoft, IBM, Salesforce, and more. Combined, Pledge signatories generate over $1.8 trillion in global annual revenues and have more than seven million employees across twenty-six industries in twenty-one countries.

Pagitsas: The large number of global companies partnering certainly indicates the importance of addressing climate change. How is Amazon meeting its Climate Pledge commitment?

Hurst: It does, and that's how we're going to make change and meet our commitment—by partnering with others, sharing information, and talking about what operational changes we're making and what we're doing at our scale.

A couple of things that underpin our Climate Pledge strategy is the transformation of our own operations, including the electrification of our package delivery fleet. For example, we placed an order for 100,000 electric delivery vehicles with an electric vehicle company called Rivian. It's a very strong signal about how we feel about the possibilities of electric vehicle technology. We've also partnered with companies all over the world such as Mahindra Electric in Asia and Mercedes-Benz in Europe to purchase additional electric delivery vehicles. We've not just said, "We hope to do this," but we're doing it. We're setting goals toward that commitment, and then we're just getting to work. In 2020, we delivered more than twenty million packages to customers in those electric delivery vehicles in North America and Europe, and we're learning quite a bit as we go.

The other thing we're doing that underpins our commitment to achieve The Climate Pledge is investing in renewable energy. In 2019, we set a very ambitious goal for ourselves to use 100 percent renewable energy by 2030 within our global operations. It's a good example of how Amazon thinks big and dives deep into how processes work to achieve those big goals. With a 2030 target date, our renewable energy commitment was set a decade before our net-zero carbon goal of 2040, which is already a decade earlier than the Paris Agreement target of 2050.

As we deployed these projects around the world, we thought we might be able to move faster. Now we've said publicly, we think we can achieve this goal by 2025—five years ahead of our initial 2030 target. We are the largest corporate buyer of renewable energy globally and already have 274 renewable energy projects underway. This includes 105 utility-scale wind and solar projects and 169 solar rooftops on Amazon facilities and stores worldwide. We went from using forty-two percent renewable energy in 2019 to sixty-five percent in 2020. We're moving fast on these big commitments and goals, including another interim goal for 2030, which is called Shipment Zero, which is our commitment to make fifty percent of all Amazon shipments net-zero carbon by 2030.

In 2020, we also launched The Climate Pledge Fund with an initial $2 billion in funding to help support visionary companies whose product and service solutions will facilitate the transition to a zero-carbon economy.

Pagitsas: What other key activities fall under Amazon's Climate Pledge commitment?

Hurst: While we're starting with actions within our own company, we also continue to make significant investments outside of our value chain to mitigate carbon emissions at scale through nature-based solutions. With this in mind, we have a $100 million fund called the Right Now Climate Fund to restore and conserve forests, wetlands, and grasslands around the world. Another example of this work is the Lowering Emissions by Accelerating Forest finance [LEAF] Coalition, an ambitious public-private initiative that Amazon helped to drive. It is designed to accelerate climate action by providing results-based finance to countries committed to protecting their tropical rainforests.

We're also using our technological size and scale within AWS to help advance climate change and sustainability research. Through a program called the Amazon Sustainability Data Initiative [ASDI], we're helping scientists and researchers worldwide access data sets like weather observations and satellite imagery by hosting them in ASDI's catalog through the AWS cloud. The data is publicly available to anyone and has already helped government officials in Africa address deforestation and gauge damage from forest fires.

We are also committed to ensuring that the people and communities supporting our entire value chain are treated with fundamental dignity and respect. We strive to ensure the products and services we provide are produced in a way that respects internationally recognized human rights.

Ultimately, our strategy is to be in service of our customers and the planet. We are going to make ambitious commitments and back them up with specific goals and urgent action. We're going to share how we're achieving those goals, what is challenging about them, how we're implementing them, and what's our progress toward them.

Pagitsas: How does Amazon quantify and share with stakeholders its progress toward its decarbonization goals?

Hurst: Amazon publishes an annual sustainability report to detail our continued path to decarbonization as we advance toward our commitment to The Climate Pledge. Achieving this goal begins with measuring our carbon footprint, identifying how to drive carbon reductions across every part of our business, and equipping our teams with the tools, knowledge, and resources to act. Our vision for the low-carbon economy of the future means that Amazon's buildings are fully powered by renewable energy, our fleets run on renewable electricity and other zero-carbon fuels, and the indirect emissions sources throughout our supply chain are zeroed out through renewable energy, energy efficiency, sustainable materials, carbon sequestration, and other carbon reduction measures.

While we are still in the early phase of decarbonizing our business, we are pleased to see meaningful progress in several areas. We will continue to rapidly scale our investments in carbon reduction solutions with large, long-term impacts that will move us forward on our path to net-zero carbon by 2040.

Pagitsas: What are customers telling you that they're expecting from Amazon around sustainability?

Hurst: We're very focused on listening to what our customers want and need, and of course, that differs around the world. We also have different types of customers. If you're talking to somebody in our AWS business, their customer will be an enterprise-level business—usually a large business we work with, but also start-ups. They may be interested in the energy efficiency of cloud-based data centers compared to on-premises data centers.

On the retail side, we have both individual customers and businesses that buy and sell on Amazon.com, respectively. In September 2020, we launched an initiative called Climate Pledge Friendly, making it easy to discover and shop for more-sustainable products. As of November 2021, over 200,000 products and more than 10,000 brands are available through Climate Pledge Friendly across beauty, wellness, apparel, electronics, household, and grocery. We partnered with trusted third-party certifications to highlight products that meet sustainability standards and help preserve the natural world.

When we began looking at how to identify products as Climate Pledge Friendly, we said, "We're not going to go in and try to tell the customer what's sustainable on our own because there's tons of work out there already to lean on." It took several years to evaluate the available sustainability product certifications in the marketplace, and we looked at over 450 or so certifications. As of November 2021, we partner with thirty-seven external certifiers—across governmental agencies, non-profits, and independent laboratories. We continue to hold a very high bar for the product certifications we select.

We also developed our own Compact by Design certification because we saw a gap to fill in the lineup of certifications. To qualify for Compact by Design, products must have best-in-class "unit efficiency." We used product attributes such as item package dimensions, item weight, and the number of units. The lower the unit efficiency value, the more efficient the design is. As an example, a concentrated laundry detergent can wash the same number of loads of laundry in a smaller volume and weight than a non-concentrated detergent.

SCS Global, a third-party certification, validation, and verification firm, analyzed our decision framework, methodology, and criteria. They provide us external validation as a credible certifying organization. Compact by Design is a great initiative that identifies products that may not always look very different but have a much more efficient design. With the removal of excess air and water, products require less packaging and become more efficient to ship. At scale, these small differences in product size and weight lead to significant carbon emission reductions. This certification and the other third-party certifications we recognize as part of this program are designed to help preserve the natural world and are meaningful to customers with a very high bar for sustainability.

These actions allow our customers to say, "I want to shop for Climate Pledge Friendly products." And it allows us to send a signal to our partners in the consumables and consumer electronics sectors that our customers want these, and we're going to continue to make it easy for them to find them. That's a good example of how we're taking what we know from a science-led perspective and translating it into something that is easy for the customer to understand.

Pagitsas: What early results are you seeing from the Climate Pledge Friendly program?

Hurst: Brands with products labeled Climate Pledge Friendly have seen higher click-through rates and higher new-to-brand purchase rates. Post-purchase customer studies show positive signals in finding the program important, switching behavior, and repeat purchase behavior.

Pagitsas: How do you partner with suppliers and sellers of products on the Amazon retail site?

Hurst: As we have done in many other areas, we want to partner with our suppliers and share what we know. Our packaging work is a good example. Over a decade ago, we developed Frustration-Free Packaging, which is a great term when you think about it. It's our acknowledgment that, as the customer, you want the item that you purchase, not all the other stuff around it. You want to get to your item quickly and with minimum environmental impact, so the packaging is 100 percent recyclable, doesn't have clamshells, and has no twist ties.

This was an opportunity to change packaging when we identified that e-commerce is different from shopping in-store. When you're in a store, you need to display the product and the marketing information for the consumer. For example, toys need to stand upright on the shelf. They need to be a certain size to attract someone's attention and have details of the benefits or special differentiating properties of that product facing the consumer. The toy may also have embedded security devices because of the physical retail environment. There's a lot that goes into designing packaging when it's stocked in a physical store.

When you get into an e-commerce environment, all that kind of goes away. You can put all the details, including customer reviews, videos, and everything else, onto the web page for the product on our site. You don't need the clamshells for security or the twist ties to keep the product upright. You can reduce all the packaging, which becomes waste when the customer receives it, and just ship them the product itself.

This example provides an enormous opportunity for us to think about sustainability differently and share that learning with our fellow competitors. Our packaging teams at Amazon developed an e-commerce packaging standard called ISTA-6 with the International Safe Transit Association. They then open-sourced it to all packaging labs used by retailers globally. We invested in some of this solution at Amazon, but it was important to be able to share the learnings and say, "We're an open-source contributor. This is good for the planet. All people need to have access to this whether you're doing e-commerce activity with Amazon or someone else."

In addition to the packaging transformation and minimizing waste, we want to make sure the product is secure. The worst thing for the environment is a customer receiving a damaged product. You have not only product waste, but you have reverse logistics of shipping the damaged good back to the supplier and making and shipping a replacement product. We want the product to come to the consumer well-packaged and damage-free. So, we also identify opportunities to ship in the product's original container, which is where we identify products that don't need an Amazon overbox. If you purchase a case of paper towels or diapers on Amazon, you do not need us to put an extra box around those. They're generally not a gift.

Pagitsas: What has been the impact of this change in shipping processes?

Hurst: Since 2015, we have reduced the weight of our outbound packaging per shipment by thirty-six percent, which is an elimination of one million tons of packaging material, which is the equivalent of over two billion shipping boxes. It's been a huge opportunity to identify how customers think about shopping differently and how we can transform our operations to take waste out of the system. You can create incredibly valuable sustainability solutions when you rethink that customer experience.

Pagitsas: It sounds like sellers are important partners in developing and executing the possibilities with sustainability solutions. Are there sellers who push back on changes to existing business processes that incorporate sustainability? What are their concerns?

Hurst: The question of pushback is an excellent one. My perspective on that is grounded in how I lead my teams, which is that I urge my team to invite all perspectives. We want to be open and transparent, and we want to hear what challenges people are having with implementing sustainability. We take those concerns and those challenges very seriously, but we break them down further to understand what's behind them. When you do that, it doesn't become a head-to-head challenge between Amazon saying, "I want you to be more sustainable" to a supplier. It's about having a dialogue with a supplier and asking, "What are your concerns?"

Generally, when we hear concerns, they center around a couple of things. There are perceptions around sustainability initiatives, such as that they might slow your business down or it might cost more to implement sustainability initiatives. There are also perceptions that some customers may want this change, but others may not want it, so it becomes important to understand the needs of different customer segments.

Pagitsas: Is there a philosophy or approach toward pushback or critique that guides Amazon as a company?

Hurst: We welcome critique. We just released a new leadership principle on this, which I think encapsulates it perfectly—Success and Scale Bring Broad Responsibility. This principle acknowledges that we're big, we impact the world, and we're far from perfect. We understand the role that we play in our communities and for the planet. We strive to do better, learn more and continue to work on things every single day.

The critics provide information to us about what they need to hear from us, which is generally more about how we're executing the strategies. What are those execution activities, not just the goals that a company sets or the commitments that they make? How is it happening? I think that we do have a responsibility, given our scale, to leave things better than how we found them. Internally, we welcome challenges at Amazon. We have leadership principles around that concept, which we call "Dive Deep" and "Invent and Simplify". We welcome inspection of our own work.

Another mechanism that guides how we operate at Amazon is to create tenets for how to work that includes the "unless you know better" option. It's a unique way of operating. When I present our work to another team, I include "unless you know better" in the document because we can always continue to improve. We can continue to integrate more science. We can continue to learn about our customers. This is all absolutely part of how we're going to do this.

Pagitsas: Climate change is a big issue, and everyone is on their own journey to understand what it is and how it impacts them. How do you and your team engage with partners at the beginning of their journey?

Hurst: People are on a learning curve around sustainability and have many questions about it. Now people have conversations around climate change at their kitchen tables. That's a big change. They're hearing about climate change in their communities that might be experiencing climate-related events, for example. Wildfires and flood events are more frequent and intense due to climate change. People feel the impact of climate change differently depending on where they live and work.

Our role is to help translate climate change issues and help answer questions like, "What can I do? What does this mean for me, my business, my products, and my services?" We want to know how we can help sellers and customers to overcome those legitimate concerns with scientifically-based information. The transformation of business toward more-sustainable practices is absolutely a competitive advantage. It's where the world is headed, and we want everyone to participate in that transformation. Yes, it's because we want to tackle climate change, but also, quite frankly, at this point, it's a disadvantage if our suppliers and customers don't understand it.

Pagitsas: What's your personal philosophy about the challenges facing sustainability at Amazon?

Hurst: I am relentlessly optimistic that we will be able to do this. It's going to take a huge amount of dedication and courage, and the ability to listen to how we need to change. And that includes all voices. This is all part of how we're going to work together. You can never have a partnership without understanding each other's side, without engaging a wide set of stakeholders. It's going to make you better at your work. It will make us better as a company, and we embrace and understand that responsibility.

Pagitsas: How have your professional experiences shaped your optimist view and influenced your leadership of sustainability at Amazon?

Hurst: I've been working in the sustainability field for over twenty years, almost approaching thirty years. I feel like I've grown up with the field and engaged with it with a wide scope. From working with the mayor of San Francisco and a United States senator, I have seen the interaction between different private sectors and the critical importance of public-private partnerships to the success of sustainability. From my many roles, I have seen what the public sector can do to set impactful policy and how the private sector, non-profit, and multilateral resources can have an impact at scale.

In my professional journey, I have watched sustainability evolve from a compliance activity into a huge value-add for the public and private sectors. I've seen how corporations can bring their scale to influence solutions for important issues we face from a social-environmental perspective. Before entering the private sector, I worked in several capacities affecting policy. I worked for BSR and led projects across different industries, which gave me a great perspective on how different industries dealt with sustainability issues. After BSR, I was the CEO of a science-based organization, The Sustainability Consortium, looking at product-level sustainability.

When I came to Amazon, it was an incredible opportunity to put all my prior experience into play and significantly impact the climate crisis. Given the industries in which the Amazon businesses operate and the scale at which we operate, we can lead through our commitment to The Climate Pledge and influence the transition of our businesses and the broader value chain to be more environmentally sound and socially responsible.

Pagitsas: Why are public-private partnerships a valuable tool for Amazon and other companies to engage in to address climate change?

Hurst: No one of us can solve the climate crisis alone. It will require partnership across multiple domains, and each of us is bringing solutions to the table and amplifying those of others. We do not have much time on climate change. Being at the table having conversations across the public and private sectors is very important for us at Amazon because we can partner with others to share information and perspective. It is a critically important part of how we're going to get this done.

If you think about what we do at Amazon, we lead with science in all our work. We model how a company can take the best available science on climate change issues—whether from recent Intergovernmental Panel on Climate Change reports to predictions on what climate change is going to do to the world our customers live in and to our businesses—and identify how we need to transform as a business.

However, we need an enabling policy environment to support the transformation. We need technological solutions to support that. We need different partners to come in to help. For example, those partners could be a non-governmental organization [NGO] that can provide information on how this issue affects their community. Or it might be a government institution at the local, state, or federal level that considers creating enabling policy environments to motivate industries to shift. A partner could also be a start-up entity that says, "We have a technological solution we need you to make a bet on for us to create to scale." The entire ecosystem of NGOs, government, and businesses needs to come together to be successful.

Pagitsas: The November 2021 Conference of the Parties [COP] to the UN Framework on Convention on Climate Change was an important global stage for public-private partnerships. Tell me about Amazon's announcements at COP.

Hurst: A great example of a public-private partnership is the LEAF Coalition. We're bringing forward other corporates to advocate for a very credible voluntary carbon market. To urgently address climate change, the solutions that are needed will require action by governments globally. Together with the governments of Norway, the United Kingdom, and the United States, Amazon and other leading companies are accelerating climate action by committing upfront to purchase verified emissions reductions from countries tackling one of the world's most critical issues—reducing tropical deforestation. During COP, the LEAF Coalition announced that it had mobilized $1 billion to protect rainforests worldwide just six months after its initial launch. We must also recognize the critical role of NGOs, Indigenous Peoples groups and local communities on the ground in stewarding and safeguarding the world's forests. We must listen to them. With the LEAF Coalition, multiple sectors are coming together to solve an incredibly complex problem.

At COP, we also helped launch the First Movers Coalition, a public-private partnership between the US State Department, through Special Presidential Envoy for Climate John Kerry, and the World Economic Forum. It works with leading climate experts and mission-aligned organizations to accelerate and scale collective impact through the purchasing power of companies globally— targeting emission reductions in historically high-emitting and hard-to-abate sectors, including aviation, ocean shipping, steel, and trucking. Unlike the green technologies available for short-distance deliveries, long-distance transport has few low-carbon technologies available.

This global, cross-sector partnership sends a clear demand signal to scale-up emerging technologies that are essential to transitioning the world's economy to net-zero carbon. As a member of the coalition, Amazon will continue to explore, test, and invest in sustainable innovations across freight, air, and ocean transport to reduce emissions on our longest routes. We're focused on bringing the right technologies to a commercial scale within the next decade, which aligns very well with our commitment to The Climate Pledge.

The First Movers Coalition is complementary to the efforts of the recently announced Cargo Owners for Zero Emission Vessels [coZEV], a new collaborative network for ambitious action to accelerate maritime shipping decarbonization. Amazon is one of the first signatories to a 2040 ambition statement facilitated by the Aspen Institute.

Another example is the recently announced addition of an Aviators Group to the Sustainable Aviation Buyers Alliance. Co-founded by Amazon Air and other top US air carriers, this group brings together organizations across the aviation sector committed to reducing their air transport emissions through investment in high-integrity sustainable aviation fuels [SAF]. We need increased production, price reduction, and continued technological innovation for low-carbon fuels to enable critical emission reductions in this sector.

Similarly, we're also focused on enabling the scaling of corporate renewable energy procurement more rapidly across more countries to help to support new investments and projects. We helped launch the US Department of State's Clean Energy Demand Initiative to encourage greater availability and affordability of corporate renewable energy options around the world.

Pagitsas: Many companies are innovating solutions to address climate change, yet they often have trouble accessing capital and scaling their solutions for commercial use. How is Amazon helping scale new solutions?

Hurst: There are several ways that we are partnering to support start-up technologies. Along with our Climate Pledge, we launched The Climate Pledge Fund with an initial $2 billion in funding to invest in companies across multiple stages of development from early-stage technologies or later developed technologies and in multiple different industries from food and agriculture to renewable energy generation, storage utilization, manufacturing material, transportation logistics, and circularity of materials. The companies we're investing in are incredible. Some are in very early development phases working with experimental technologies. We come in and say, "We think this is a solution that will help to decarbonize our own operations at Amazon. We want to make a bet on scaling this solution or technology."

An example of an early-stage solution is CarbonCure Technologies, which is commercialized lower-carbon concrete. In CarbonCure's process, recycled carbon dioxide is injected into fresh concrete. We can use that product in new construction. We're using it to build our second headquarters in Virginia.

We also have invested in electrofuels solutions through Infinium. The company focuses on converting carbon dioxide and hydrogen feedstocks into net-zero carbon fuels for use in today's air transport, marine freight, and heavy truck fleets. We've invested in Turntide Technologies, which develops efficient motor technologies that reduce the energy consumed by heating, ventilation, and air conditioning in buildings. We're utilizing these technologies within our buildings and evaluating how they deploy. We ask ourselves questions such as, "Was there a significant reduction in electricity usage?" followed by "How do we help them to continue to build a better product?" We've also invested in a company called ZeroAvia, which is a leader in zero-emission aviation, focused on developing hydrogen-electric aviation solutions.

We're looking at not just what we need as a company to decarbonize our own operations and what would be beneficial for others, we're also testing and trialing and utilizing these types of things within our own company and giving feedback and information. It's an interesting process to go out there and find things that we think could be significantly impactful. We not only put financial capital into them to help scale, but we also have major global operations that allow us to implement this technology early on and we can see what the results of that are.

Pagitsas: The scope of sustainability at Amazon is quite large, with a team of many experienced individuals and significant capital for projects and investments. What leadership advice do you recommend to a new CSO who is initiating their company's sustainability strategy and may not have the same resources or buy-in?

Hurst: My number one recommendation is to be open and transparent. Be a seeker of all the best ideas. Dive in to fully understand the business you're in and the challenges your peers face, and identify where sustainability might provide solutions for the business. Just get started on being a partner to whatever type of business you're in.

When I came into Amazon seven years ago, we had a retail business, a cloud computing business, a studio business, private brand products, and large fulfillment and logistics operations. I needed to understand the needs of the person who was running one of our fulfillment centers. What were their pain points, and how could sustainability help resolve some of those?

In the case of AWS, how do we help our customers understand how much more energy-efficient it would be to move to the cloud? Our Amazon Studios and Prime Video businesses are an opportunity to communicate about sustainability to customers who interact with our movies and TV shows through our content creation. There's a lot of different ways in which you can address sustainability issues. It starts with knowing your business and the opportunities to implement sustainability.

Also, just get started and be open to transforming, pivoting, and developing your ideas along the way when your peers come back to you and say, "This is working, this isn't working, here's what we're seeing, here's what we learned." We are customer-obsessed at Amazon, and so it is a great leadership principle to lean in on. My message to my team is, "All of those internal partners, peers, business leads are our customers. We need to fully understand what they need," and go from there and give some early wins to the business on sustainability. It's important for the business leads to hear, "We can cut costs, we can save time, and we can have a better customer experience through some of these sustainability initiatives." This approach helps the business leads see how sustainability could grow within Amazon and their own business. Bringing in that business and customer focus is the best way to get started.

Pagitsas: Does "getting started" mean everybody internally is on board from Day One?

Hurst: The way I would think about that is you don't necessarily need to have everybody on board. I was very lucky at Amazon that our senior leadership understood sustainability and its importance to our customers and our business. I had that coming in the door. However, there are many things our business leaders need to think about on any given day.

Pagitsas: At Amazon, what are those competing responsibilities or targets on business leaders' plates?

Hurst: First and foremost, there's the safety of our own people and of anyone working for Amazon, with Amazon, and in our communities. Second, we need to look at cost, quality, and continuity of product. At any one time, there's a lot of different things on any business leader's agenda. However, I want sustainability to be on the priority list as well. I want it to be a factor that's considered in any business decision-making. It may not always be the top of the list, it may not be the number one thing that drives the decision, but it should be a consideration.

Pagitsas: How do you influence your peers to at least evaluate the sustainability factors of a business decision?

Hurst: At Amazon, we need to give people information and data. We need to be able to tell them the sustainability impact of the decision they're making for them to put that into the mix as they evaluate trade-offs.

Sometimes, we make trade-offs in favor of sustainability that can also have great customer outcomes and deliver cost-savings, or it can be cost-spending, and we decide to spend more rather than less. Having the information to understand the sustainability impact of the decision you're making and putting it in the business case analysis is incredibly important. It then becomes integrated into the business decision-making process, which is essential to achieve sustainability goals in partnership with the business.

Pagitsas: Is there a key question you ask yourself and your team to consider when partnering with business leads?

Hurst: Yes. It is, "How am I allowing and enabling and empowering all of our business leaders to think about sustainability as they continue to deliver great products and services for our customers?"

Pagitsas: My takeaway is that you're empowering business leads to drive and make the right sustainability decisions, but at times, you may not have all the sustainable outcomes you want.

Hurst: Yes. It's a realistic way to look at it. It's pragmatic, but I also think it's because when people can make a better and more-sustainable decision, they want to do that. They want to understand the data around sustainability. I work with people who are highly interested, motivated, and excited about sustainability, so I want to give them the information to make that more-sustainable decision. If they don't have the data, they can't do that in the moment.

The other critical point about this approach is that Amazon is a large and complex company. There's no way that my central sustainability team can discuss every business decision with every business lead. We drive governance of sustainability methodologies and science-driven decision-making, and we drive a lot of the data. Still, we're not going to be at every table where business decisions are made.

Pagitsas: This is a particularly important point for sustainability leaders at large companies where the number of business leads will far outweigh the number of people assigned to the sustainability team.

Hurst: Yes, absolutely. We must be able to build a system that embeds sustainability principles and a mechanism to execute on those principles in a way that grows with our business. That is the key to successful sustainability. I want our team's ultimate goal to be that we've embedded sustainability thinking so deeply within the company that it is just intrinsic to how we do everything.

I'm excited about the fact that we seem to be in a new era of business leaders saying, "We understand we have to make a commitment. We want to be courageous about making that commitment. We don't know all the ways that we're going to get there, but we're going to start investing in new technologies. We're going to start to think about how to revise our algorithms internally to optimize the way that we operate and make big changes." These may not be the most visible things to most people, but they are hugely impactful because we now have the data to understand where the changes need to be made and are sharing what we learn. When you think about those hundreds or maybe thousands of companies that are committed to doing this over the next couple of decades, the impact of our partnerships gets accelerated.

Pagitsas: I agree with you on the importance of distributing ownership of sustainability and trusting people to balance sustainability and business goals. Finally, what other ideas are central to your approach to leading in sustainability?

Hurst: There are a couple of additional points. First, I absolutely want the best ideas to come and flow freely into sustainability. I want the best talent to flow freely into our team, and I also want to export that talent to all other teams at Amazon.

Second, it's important to ensure that my team has a very deep mix of perspectives and of the different lenses that people are looking through. I love a healthy debate around sustainability. That makes us so much better as a team and as a company when we can encourage that debate, when we're able to engage in it, and when someone can change my mind.

Third, ruthlessly and relentlessly prioritizing sustainability solutions for the company is one of the superpowers that our team brings. We need to constantly focus on the biggest levers we can pull for Amazon to tackle sustainability challenges globally given our scope of operations and complex businesses.

Lastly, I hope any leader I work with would say that I trust and respect them professionally and personally to integrate whatever they need to do in life and work. This is a challenging area to work in, and people bring a lot of passion to sustainability. We want that passion, and I want them to have long and successful careers in Amazon. As a leader, I value people's ability to balance and integrate their life and work and bring their whole self. It makes all of our work better in this area.

Steve Varley

Global Vice Chair — Sustainability
EY

Steve Varley is EY Global Vice Chair — Sustainability. EY is a global professional services organization with more than 312,000 people worldwide and $40 billion in revenue for the financial year ending in June 2021. Steve leads EY's climate change and sustainability strategy globally and helps clients create business value from sustainability and accelerate their transition to a lower-carbon future. Steve is founding co-chair of the S30, a group of more than 30 chief sustainability officers (CSOs) from some of the world's leading businesses. It launched in 2020 as part of His Royal Highness The Prince of Wales' Sustainable Markets Initiative.

Steve was previously EY United Kingdom and Ireland (UK&I) Regional Managing Partner and EY UK Chair. He joined EY in 2005 and has nearly 30 years of client and consulting business experience in various sectors, including pharmaceuticals, oil and gas, and public services.

Steve was a member of the UK Prime Minister's Business Advisory Group for both Prime Minister David Cameron and Prime Minister Theresa May and was also a UK government business ambassador for the professional services industry. Steve is a founding member of the Social Business Trust and a UK member of the International Chamber of Commerce governing body, the Chairman's Advisory Group for the British Museum, and the Productivity Leadership Group. Having earned a BEng at Loughborough University, he is now on its board.

© Chrissa Pagitsas 2022
C. Pagitsas, *Chief Sustainability Officers At Work*,
https://doi.org/10.1007/978-1-4842-7866-6_9

Chrissa Pagitsas: Let's tackle your career first, Steve, before we touch on EY's sustainability strategy and its work with clients. What was your professional journey to becoming EY's sustainability officer?

Steve Varley: I began this role on July 1, 2020, when I became EY's first-ever global vice chair for sustainability, reporting directly to our global chairman and CEO, Carmine Di Sibio. It is the first time we've had this role at EY. Its creation was a reaction to significant client demand, our own people's interest, and a recognition that we at EY could do a lot more if we enshrined our thinking and gave accountability to a senior leader.

Before I took on the role as the global vice chair for sustainability, I spent nine years leading EY's business in the United Kingdom and Ireland. As a strategy consultant, I'd like to think that I help our clients solve their most complex problems. That's what I set myself out to do. I'm an engineer by degree and training, not an accountant or an auditor.

Pagitsas: And not an environmental engineer, correct?

Varley: Correct, my degree is in civil and structural engineering. My particular skill sets revolve around growing businesses, solving complex problems, and driving value from both of those. As you say, I'm not a sustainability expert, but I have a lot of people around me that are, and, together, I think we can make a really big difference.

Part of my work now is to bring business as a powerful positive force to the sustainability agenda with stakeholders so that we can create financial value together and value for society and the planet. That's what I think I can bring, a business perspective. I ask my team, "How do you create growth? How do you create value? How do you solve problems?" Together with help from my sustainability colleagues and the rest of the EY teams in assurance, tax, and corporate finance, hopefully, I can make a big difference as the first EY global vice chair in this area.

Pagitsas: How did EY's expanded focus on sustainability come about as well as your new role as global vice chair of sustainability?

Varley: Great question. In 2018, EY began a refresh of its global strategy, which became known as NextWave. I was involved in the working groups that were creating NextWave. As you might imagine, we held workshops around the world—London, New York, and other places to craft this new global strategy to power us forward.

Every time the NextWave strategy came to me for discussion, I pushed into it what we'd now call "green agenda" items or "ESG" and "sustainability" items. It felt right to me based on my own view but also on what my most progressive clients were telling me and what I was learning through meeting commentators and experts globally. They, too, viewed sustainability as a source of growth and value for business and that topics like climate change and social justice should be pushed into the heart of the NextWave strategy. It felt natural to me—and the right thing to do.

Therefore, after nine years of running our United Kingdom business, it felt very natural to discuss with my global chairman and CEO how I could help make sustainability an even bigger focus for EY and bring more of a strategy lens to it to help clients make change happen more quickly and successfully.

Pagitsas: It seems EY's growth strategy, as captured in the NextWave strategy, is tightly linked with sustainability.

Varley: Yes, it is. I was emboldened in it being the right approach because I saw the energy for my fellow partners in the working group when we talked about the power of sustainability to get the organization to the next level. For the partners, sustainability will get us to the next level of helping our clients be even more successful and helping EY bring its purpose of building a better working world to life across the globe. It is critical because EY has such a large opportunity to work with clients to deliver a more sustainable future, and our 300,000-plus people can also contribute here. As we shared the early thinking on the NextWave strategy with our people, I saw their energy levels pick up as well.

When I bounced the idea with my clients, I saw them lean in and express that they were trying to do the same thing with their organization or that they'd like to. I got many points of validation which is, I think, always important in any new journey. The drivers for clients are now over and above moral and ethical reasons for doing good. The most progressive clients are embedding sustainability into their strategy and using it to drive growth and value for multiple stakeholders, including the planet, their people, and their investors.

Pagitsas: How did you and your team identify the boundaries of EY's own climate strategy?

Varley: As a sustainability team working with our highest governance body, the Global Executive, we reflected on the sustainability and ESG definitions first. We went back to that great foundation of the United Nation's seventeen Sustainable Development Goals [SDGs]. We looked at what we did naturally against the seventeen goals, and then we looked at where we thought we could contribute even more. While we would say we connect with and have a role to play with many of them, the one that we felt gave us energy and was a bigger and bigger issue for our clients was number thirteen of the UN SDGs, which is climate action.

We then asked ourselves the question, "What does leadership in climate action look like?" We looked at our peer group in professional services, looked at the world's leading companies and the targets and narratives they were using, and how we connected with their purpose.

Pagitsas: What public commitments has EY made toward the UN SDG Climate Action goal?

Varley: The carbon ambition commitment was launched on January 25, 2021. Our primary goal is to become carbon negative and remain so by removing more carbon than we emit. Our immediate goal is to become net zero by 2025 by reducing our absolute emissions by forty percent from a 2019 baseline. To ensure we are on the right path with our projections, we had all our numbers certified by the prestigious Science-Based Target initiative [SBTi], following a good dialogue with the organization a couple of months ago.

We've got a seven-point execution plan to deliver this ambition. The seven points are focused on our own operations, such as moving to 100 percent renewable energy and reducing our air travel, and on working with our suppliers to help them reduce their own emissions. As we're a simple business, we can—and we should—go carbon negative. We're simple because we don't manufacture anything or have a complex business operation. We are a people business powered by technology. We fully appreciate that lots of other businesses are a lot more complex, especially in the industrial world. In comparison to EY, those businesses need to account for the greenhouse gases emitted in their manufacturing facilities or outbound supply chains.

Still, we hope that more and more businesses that can go carbon negative will go carbon negative. We hope this will create a more positive connection between business and society, as society increasingly recognizes that we want to be part of the solution to a warming planet.

Pagitsas: How do you integrate sustainability into business practices at EY and for clients?

Varley: This is one of my favorite discussions. How do you make organizations change? Or how do you help organizations change for the good? I say to clients, "A lot of us know how to make companies change. But do you have an approach that creates purposeful change time and time again and change that ends up being good for people and the planet?" Because not all change has been good.

We asked ourselves the question, "How do we at EY create purposeful, positive change on this agenda?" There are many ways to do it, but the biggest lever we have is our culture. Culturally, EY people are built to solve clients' problems. We also find tremendous value in playing back into the organization and sharing the stories of clients who want help on these agendas. Part of the approach we've applied over the last eighteen months is to invite our leading clients into the organization to share their stories of where they need help on their environmental and social agendas.

Then EY's problem-solving DNA kicks in. If you create other stimulants, you tend to find that the organization starts to respond to those client issues, opportunities, and challenges with a suite of solutions. Then we pilot and test them with some clients. When the solutions are ready, we industrialize them. That's one of the ways you make good change happen at EY time and time again. Now, there's more to it than that, but it becomes more challenging if you don't get that bit right.

Pagitsas: What services does EY provide to support the sustainability strategies of clients across industries?

Varley: First, there's no doubt that the climate agenda is creating a significant regulatory reporting burden for many companies. There's an explosion of sustainability or environmental, social, and governance reporting coming, not just with the European Union's Corporate Sustainability Reporting Directive, but also from the SEC [the US Securities and Exchange Commission] and major countries like Japan and China. There's a lot more reporting obligation. We at EY have a heritage in assurance, auditing, and reporting. Supporting our clients' reporting needs is one key solution we offer.

Other things that we've been doing over the past few years on this agenda includes helping clients identify or expand business opportunities. For example, we have a good business in helping clients explore and then invest in renewable energy. We've helped some major industrials invest in business units to create wind power on the northeast coast of the United States and off the coast of Europe.

We help the companies create joint ventures, do the business case, secure financing, and navigate any regulatory hurdles. We help them source the components and build the new joint venture organization structure that will look after the wind farms in some very difficult environmental conditions, sometimes hundreds of miles off the coast of major markets. We help companies create financial instruments such as green bonds to help them finance their own transition. Green bonds are a fixed income debt instrument like any other bond issued by governments, supranational organizations, and companies. Unlike regular bonds, the funds raised are allocated to finance environmental and climate projects. We help banks work through their loan books to understand the financed emissions they have within their portfolio and help them engage with their clients on how to reduce their greenhouse gas footprint.

I mentioned upcoming regulations in many jurisdictions. There's also increased taxes, not just on carbon dioxide. China has just created the largest emission trading system in the world. The European Union has a quite mature carbon trading emission system. After exiting from the European Union, the United Kingdom has now created its own system. There's a whole debate in the European Union on carbon border-adjusted taxes and what that would mean. In addition to taxes on carbon and climate, many countries around the world have now deployed taxes on plastics. We help our clients make sure that they understand and obey the law in these countries on plastic taxes.

Additionally, we help businesses make decisions on their supply chains and where to build their manufacturing plants. We also help major industrials such as steel manufacturers on a more practical level think through how they can decarbonize their business by exploring new technologies. This may range from exploring what hydrogen's role could be as a heat source to how they can access more renewable electricity.

We are helping several clients explore the possibilities over the longer term of decarbonizing their logistics and transportation operations. That could either be by shifting to electric vehicles or hydrogen vehicles or both. Those projects create business cases that require hundreds of millions of dollars of financing and change management programs that we should expect to take at least five years because the supply side of EV [electric vehicle] or hydrogen vehicles is so much in its infancy. Clients will have very long horizons on some of these major change programs. We're helping clients, especially major industrials, think through what role emerging technologies like carbon capture usage and storage could play. In parallel to advising our business clients, we're advising governments on how they create regulatory and incentive frameworks to encourage these technologies.

Lastly, maybe the biggest area with which EY can help all clients is to complement their sustainability strategy with a strategy on change management. Once they've established targets, how do they get their organization to change for the good to meet those targets? EY is at its best when we're collaborating with clients to make change happen. That's a big part of what we do, including some interesting work that we've pioneered using the nudge theory from Richard Thaler, the Nobel Prize laureate. He writes about how small changes to an ecosystem can adjust behaviors to create big positive outcomes in his book *Nudge: Improving Decisions About Health, Wealth, and Happiness* [by Richard H. Thaler and Cass R. Sunstein (Penguin Books, 2009).] That's been fascinating as a journey.

Pagitsas: Clearly CSOs lead both change management efforts and the delivery of environmental and social outcomes through their company's core business services, products, and services. What is an example of a sector that is grappling with major change?

Varley: There are challenges for our banking clients and the overall banking industry in funding the transition of major emitters. Banks want their clients— typically major industrials and extractives—to use bank services such as financing while encouraging them to reduce their greenhouse gas emissions over time. We help our banking clients address how they can do that. How do they do so while ensuring that there's a carrot as well as a stick? How do they train their workforce to position what could be an entirely new set of products, green bonds, and other financial instruments?

These are huge change management issues to address, especially as banks are trying to work out the scope of their emissions resulting from their financing transactions. Many of them are starting to understand the enormity of financing the transition from black to brown to green industries in a way that both allows continued economic growth and provides industries with the financial capital needed for this huge pivot or metamorphosis for their businesses.

Pagitsas: In addition to change management, an additional theme seems to be that clients need to dig deep into the supply chain and expect a longer time horizon to create the change. Is there a larger, systemic change the broader business community is experiencing?

Varley: Yes. I think there's an underlying shift in how businesses define and create longer-term value. There is even more tension within organizations that run their businesses by the day, the week, the quarter, and what that means in the context of climate change. Unless we take a long-term view, we won't deliver a sustainable planet or sustainable businesses because we need to make changes that benefit the next generation but which, frankly, many business leaders today may not get to personally experience during their careers.

Pagitsas: The definition of value we're talking about here is whether businesses have a role in creating environmental and social value while also creating financial value. Are your clients still asked this question, Steve? If so, how do you respond, or how do you help them respond? What are the real questions that should be asked about sustainability?

Varley: I think there is still a good, healthy debate about the connection between value and sustainability. In my mind, it's a virtuous circle where if you can create more value from sustainability, you are compelled to become more sustainable, and then you can create even more value. That's an optimistic way of thinking about the situation, but one that I found unlocks some people's minds as they explore what their business can do on this agenda. There is value in there. However, sometimes it does take more effort to find or unearth the value.

It also requires a bit of a reframe on who you're creating value for. Many people quote Milton Friedman, who said, "The business of business is business." I've played with that in some media interviews and said, "The business of business is *bigger* than business." We have more to contribute than business, which gets you into the deeper conversations. If the business of business is bigger than business, it means you impact society. You impact the planet.

I think capitalism has many faults. However, Winston Churchill said about democracy, "It has been said that democracy is the worst form of government except for all those other forms that have been tried from time to time."

Similarly, I believe capitalism isn't a perfect system. But it is a great force for positive change when deployed correctly by people with the right values and with the right framework around them. Sometimes, it can also be very destructive.

However, if we can harness the power of capitalism, which is about creating value, if we can get the lens to be not just on financial value, but also on society, people, planet, if we can make that happen and then spin the virtuous circle of creating value for sustainability, we get to engage this fantastic force for good called modern-day capitalism. That's fundamental to a lot of thinking we have within EY. We are trying to help clients create a range of "values." Some of them may be financial value, planetary value, or societal value. The language, I think, is the same.

I have an optimistic view that financial value can and does flow if you do good for the planet and society over the long term. If you can get them aligned, you can create powerful organizations that create value for multiple stakeholders. I hope it makes the world more sustainable too.

Pagitsas: Let's talk about the intersection of sustainability and executive leadership next. Do you see that the skills you apply in your role are the same, overlapping, or different from those you used as the head of the United Kingdom and Ireland business?

Varley: Some things are the same, such as the ability to convene a group and collaborate with them to find a better solution. I think that's a part of any leader's role. There's a great quote from Napoleon Bonaparte. Although it might be an odd choice for an Englishman to quote that particular French military figure, I like it a lot. Napoleon said, "The job of a leader is to define reality and give hope." I've used that quotation a lot in my leadership roles in the past. The first thing you do as a leader of a group is make sure we all see the same reality. Once you have that, you have an obligation to create a hopeful future.

That leads me to what's different about the CSO role than the ones I have had in the past. The context that sustainability is operating in is very different. The context is very different because all of society is impacted by climate issues—from indigenous tribes in island states to urbanites living in cities next to major rivers, with both groups concerned with increasing sea levels. We ask ourselves, "Is this sustainability wave a similar wave of change to digital?" The digital transformation of a company was quite hidden to many of their own consumers and, therefore, much of society. While there are some similarities, I think the biggest difference is society's role in driving change. The digital transformation hit lots of companies probably twice as hard as we expected. However, I don't remember society at large expressing a need for companies to "hurry up and digitize."

On sustainability, however, I hear a huge cry from society and governments in a somewhat challenging way because they see businesses as part of the problem. For businesses to decarbonize more quickly and do more than meet the basics of the Paris Agreement is very different from the expectations from businesses engaging in the digital transformation, which was arguably a more straightforward ask. The ability to listen more acutely to external stakeholders and hear society's voice is something new for many of us as leaders. It really does stretch your leadership boundaries to be more inclusive and thoughtful of, as people call it, the externalities of your business operations. That's what I'd say is one of the differences for me.

Pagitsas: What are some of those externalities you're now hearing?

Varley: The most common externality is greenhouse gas emissions. You also have pollution and air quality impacts. Water preservation is another. Lastly, there's the impact on biodiversity. In this next phase of capitalism, we must price in these externalities so that business can be held to account for its wider operational impact.

Pagitsas: For a company to successfully apply a sustainability strategy, do its leaders have to believe that business should deliver social and environmental benefits?

Varley: Yes, I think they have to. My position would be that one of the leadership dialogues over the last years, if not decades, is authenticity in leadership. As I've been on my personal leadership journey, I've learned time and time again that the best leaders are the most authentic ones, not the ones that stand up and have their suit of shiny armor that can deflect every question. It's the ones that evidence their vulnerabilities, their anxieties.

I go back to the quote from my friend Napoleon about establishing the reality and giving hope. I think a leader has a role in giving hope in the face of tough challenges, but they can do that and still be very authentic. There has to be a genuine understanding of the situation that we're all in. You then have permission to engage people and take them on the journey with you—be they your employees, customers, or end consumers. Leaders do have to believe in sustainability because it's an authenticity test. They need to be authentic to bring stakeholders with them.

I've got a view of life that human beings have several tests they do. It's subconsciously deployed when they meet individuals, and one of them is whether they find them to be authentic or not. We all know what it's like when you're working with someone who's not authentic. On this agenda, in particular, I think leaders have to be authentic.

Pagitsas: I appreciate the point on authenticity. Some of the sustainability issues link to very personal issues for employees. To hear a leader talk about racial inequity, for example, but to perceive the leader as being inauthentic undermines the leader's credibility and, therefore, the success of the overall strategy.

Let's come to the last question of our interview. What are you optimistic about in relation to sustainability and business?

Varley: My frame on work and life is that I'm an optimist, but I'm an optimist that worries a lot. I think more and more of us are waking up to the challenges and the perilous position that we collectively put our planet in. I feel from the business community that there is a wholehearted understanding that we need to be part of the solution. Frankly, if we are not, I think we will be judged by society, our employees, our consumers, and our families to have not acted in the right way during these perilous times.

One of the challenges we have on climate change is if we don't deliver on reducing the planet's temperature increase, I think there'll be societal breakdowns as many populations will be displaced. They won't be able to farm in the same way or make a living in the same way. Entire ways of life are at risk. There's double jeopardy if we as business don't quickly become part of the solution and learn to collaborate on a new way of doing things. Not only will our planet heat up, but some of those adjacent issues will also exponentially grow in their impact, including social inequity and unrest. If we don't come together to solve the climate crisis and don't make this, as they say, "the decade of action," I worry these other adjacent points will impact us even more.

If I think about the future, I'm optimistic because I do think that more and more businesses are leaning in to become part of the solution. I'm finding businesses are willing to collaborate, even with competitors, to be part of the solution. I think there will be innovations and new ways of partnering, especially new ways to collaborate. In this new world, we'll emerge having averted the climate crisis, and we'll have created a fairer society. We can use those learnings to go after the range of adjacent challenges that are nearly as pressing as climate change, such as plastics, biodiversity, and social equity.

Sophia Mendelsohn

Chief Sustainability Officer and Head of ESG Cognizant

Sophia Mendelsohn is chief sustainability officer and head of environmental, social, and governance (ESG) at Cognizant. Cognizant is a multinational provider of information technology, consulting, and business process services. It had $16.7 billion in revenue and ranked 197 in the Fortune 500 in 2020. Sophia leads the company's strategy to integrate ESG into its purpose, strategy, systems, and technology solutions.

Previously, Sophia was the first chief sustainability officer at JetBlue Airways, where she shaped JetBlue and the aviation industry by pioneering airline carbon neutrality and sustainable aviation fuel use. Prior career activities include serving as head of sustainability, emerging markets for Haworth, a multinational manufacturer in the corporate real estate industry, and working for the Jane Goodall Institute in Shanghai, establishing environment programs in offices and schools in China.

Sophia has been recognized for her leadership and commitment to corporate citizenship. In 2016, she won the US Environmental Protection Agency's Climate Leadership Award and was named Climate Leader of the Year by Ethical Corporation. In 2020, Bloomberg named her one of the top 30 people leading the climate charge.

© Chrissa Pagitsas 2022
C. Pagitsas, *Chief Sustainability Officers At Work*,
https://doi.org/10.1007/978-1-4842-7866-6_10

Pagitsas: Let's start by discussing sustainability as a concept. What are its origins and its relationship to business strategy today?

Mendelsohn: Historically, sustainability was not seen as necessary to guide strategy. Born out of corporate philanthropic roots, charitable contributions and small actions were the first stage of corporate sustainability. From there, it evolved to its second stage—a marketing and reputation strategy. And then again from there, it grew into something that CEOs and boards of directors largely saw as a means to protect revenue but not *generate new* revenue. As sustainability has become an increasingly urgent and pivotal topic, you'll be increasingly hard-pressed to find a CEO that would tell their shareholders and their board, "Sustainability is just a marketing or PR [public relations] strategy."

Now that we know where it came from, we need to explore where sustainability, now called ESG, must go next—which is, be a part of the business strategy. Strategy generates revenue, and revenue must generate profitability. ESG must overcome the idea that it is tangential or peripheral to a business' true strategy. The "doing well by doing good" phrase encompasses the idea that you can achieve two goals at the same time—one, achieve traditional business objectives, and two, appeal to non-financial stakeholders. That is the baseline for modern and contemporary ESG.

Pagitsas: The phrases "doing good by doing well" and "doing the right thing" are indeed popular in the business vernacular these days. Is it meaningful for businesses to use these phrases?

Mendelsohn: I generally shy away from the "doing good" language because we all define "good" and "right" differently. What is important to recognize from this language is the bifurcation of impacted stakeholders from profitability. That is, you could make money, *or* you could be "good."

In her book *Reimagining Capitalism in a World on Fire* [(PublicAffairs, 2020)], Rebecca Henderson shares that the first dean of the Harvard Business School defined the school's purpose as educating leaders "who would 'make a decent profit, decently.'" It is relatively recently that we've seen business come back to being a decent, in terms of considering impacts to multiple stakeholders, and sustainable profit generator. In the past twenty years, the reintegration of those two concepts has been led by some notable CEOs and external academics and thinkers. Paul Polman, [CEO of Unilever from 2009 to 2019], is an obvious one. Under Paul, Unilever embraced the idea that a brand could deliver social benefit through its products.

Pagitsas: Can you pinpoint when the two concepts' reintegration was noticed by global businesses?

Mendelsohn: It should be noted in history that there was a shift when we hit the 2019 World Economic Forum held in Davos. I created a phrase and frequently use it—"What a difference a Davos makes." The shift came to public companies first.

Large public companies said, "We've been doing sustainability for years!" And the institutional investors said, "Yes. We've been reading your sustainability, CSR [corporate social responsibility], citizenship responsibility, and ESG reports for years. But we still can't get good data out of them. Whatever you've been doing for years is not working, so now we're going to ask you to report the same way we ask companies to report on their financials—systematically and numerically. We're going to do make sure what you report is measurable and comparable. We're going to do it with goals. We're going to ask for public disclosures that we'll have you treat with the same rigor as public filings. In short, we're going to start counting carbon the way we count money."

The business leaders poised to take advantage of that moment were the leaders who had been saying, "Sustainability is also about money. This is about financial performance." And they did that in a number of ways, including by launching ETFs [exchange traded funds], issuing green bonds, and discussing climate's impact during earnings calls.

Pagitsas: Why did institutional investors make this shift in 2019 around transparency, reporting, and awareness of the impact of climate change on the public companies they were investing in?

Mendelsohn: I think what shifted for institutional shareholders was their frustration with growing, unabated risk. Social and climate risk that had been labeled long term and had shifted to medium term was now current. Yet, most public companies were not talking about these topics in earnings calls or financial filings. Investors started asking themselves, who will get left holding the bag when climate change threatens investments? Nobody wants stranded assets on their books.

You couple that with social unrest, a widening skills gap, and labor shortage, all which impact profitability and restrict revenue. This, in turn, can lead to social unrest and destabilize society, which then destabilizes the economy and further deteriorate profitability and revenue. Collectively, these issues represent major risks, and around 2019, there was a collective groundswell of agreement that stakeholders should start addressing the problem.

Pagitsas: What is the scope of the strategies that you led at JetBlue and now at Cognizant? They are in two very different industries, the airline industry and the IT and business services industry, respectively.

Mendelsohn: When I was at JetBlue from 2013 to 2020, we moved the needle on ESG and sustainable operations. We took the concept from being an internal communications approach to being a core business risk mitigator and differentiator. We paved the way in the industry for sustainability-linked loans, and sustainable aviation fuel. We thought about how the customer feels when they're with the brand—safe, on time, cared for, comfortable. I saw around the corner that social and environmental responsibility would be part of what the customer would want to feel when flying with us. I integrated it into what the customer felt when they were with the brand and into why an inflight crewmember or support center crewmember, which I myself was, would want to be with the brand.

As with JetBlue, I was brought on by Cognizant to start up the mainstreaming or professionalizing of their ESG function. This time, however, the market was much more mature in what it was asking and looking for. I joined Cognizant in 2020, after the turning events of the World Economic Forum held at Davos 2019, when climate change and ESG burst into the business mainstream in a new way, and in the middle of COVID-19.

Cognizant is going from zero to sixty on ESG. We have a new transformation CEO, Brian Humphries, who is taking a great company and helping it adjust to a post-COVID-19 world. I always say that Brian was already giving ESG at Cognizant a leg up when he established our company's purpose "to improve everyday life" before he hired me. It gave me a North Star to speak to in my strategy.

Pagitsas: What projects are you and your team working on that will propel Cognizant's sustainability strategy toward the North Star?

Mendelsohn: We are a people-based business, which means we understand the importance of attracting and retaining talent. Our overall ESG vision is to use our technologies, knowledge, and partnerships to engineer new levels of environmental and social benefit for our clients and communities.

While there are many important ESG factors, our core focus areas are generally climate and diversity—and the reporting out of our progress on these topics in decision-useful ways. We are creating and executing a path to net-zero emissions—an example of the *E* of ESG. We are linking our social impact programs with our hiring pipeline—an example of the *S*. We are reporting this to investors with the same seriousness we do our traditional business metrics—an example of the *G*.

Pagitsas: Are greenhouse gas emissions reductions a priority for Cognizant?

Mendelsohn: As a service company, our business doesn't emit a lot of greenhouse gas emissions because we do not have manufacturing or logistics. However, how few emissions is few enough? I think the market's very clear response is any more than zero is too much. I acknowledge that our emissions footprint is comparatively small to even light manufacturing, but they're still our emissions, which makes them our problem and our opportunity.

The other thing that matters when it comes to ESG for a company with relatively low emissions is that while you might not have a lot of emissions, you have a lot of clients. And your clients have a mission to reduce their greenhouse gas emissions. And your clients have the same shareholders. If shareholders are asking you about your emissions, guess who else they're asking? They are also asking your clients. So, it's not only the number of emissions we have, but it's the number of technology tools we have in the toolbox to help other people reduce their emissions.

Pagitsas: Other companies in the market, like consulting or marketing firms, have relatively light emissions footprints. Would you argue that they too should focus on zero emissions?

Mendelsohn: Leaders in companies that might not have a big public-facing brand or do not have a large footprint like an airline may not feel the immediate pressure. However, you still have stakeholder pressure. Even if Cognizant emitted only one car's worth of greenhouse gases, I'd still have nearly 300,000 colleagues who would be asking what we're doing about it. We are on a mission. Almost any emissions are too much.

Pagitsas: Given your experience starting up ESG at Cognizant and JetBlue, what advice would you give someone starting up the ESG function at their company?

Mendelsohn: Within ESG, you first need to start up ESG functions that provide governance oversight and a controls environment over what will become your key public ESG data, goals, and claims. Then within starting up ESG functions, it's working with top management and the board to make sure the bases are covered. You're establishing milestones that don't get fuzzy and infrastructure for strong oversight.

Pagitsas: As a senior executive within Cognizant, how do you partner with the board, the management team, and your C-suite peers to achieve success?

Mendelsohn: Cognizant's board of directors seeks to ensure we are integrating ESG into the existing enterprise priority matrix. Our governance and sustainability committee, whose charter we reconsidered and relaunched in 2021 to emphasize sustainability and ESG, actively engages in the oversight of ESG's enhancement and integration at Cognizant. And of course, I work closely with our general counsel to give the company a point of view on specific ESG risks and opportunities. Based on this guidance, the CEO, the CFO, and the board of directors can determine how they want to transform the company to engage in a low-carbon, inclusive economy.

Pagitsas: The strategies and projects that you have implemented at other companies and are implementing at Cognizant are far-reaching with long-term economic, social, and environmental goals. What drives you and other sustainability leaders to commit yourselves to these ambitious strategies?

Mendelsohn: I have been offered traditional commercial roles, all of which I turned down, to stay in ESG long before we knew what ESG would be today. I think ESG officers who have been in the role much longer than 2019 felt like we were watching a revolution in slow motion. We knew the way markets were disregarding externalities, like carbon, could absolutely not stay the same. The system of capitalism lifts people up. It alleviates poverty, it creates opportunities. However, capitalism would need to shift and change from the inside to address sustainability problems, including the externality of carbon and the lack of investment in human capital. Pressures from NGOs [non-governmental organizations] and other external pressures have a role in the ecosystem, but they aren't going to move the needle alone.

I'll tell you my personal story of that moment of realization. I lived and worked in China for a number of years. I was speaking to a Chinese friend who said to me, "You know, my traditional hometown is a manufacturing center. And now everyone is talking about cancer and consequences of pollution." Here is someone who would 100 times over choose those manufacturing jobs, a thousand times over would put those factories and those manufacturing jobs in his home community. He just wants to do it without the negative side effect manifesting itself in his community.

Now in the traditional dynamic, the traditional business voice would say, "Ah, but how much more does it cost? Who's going to pay? Because I tell you that the consumer isn't going to pay." It's the folks who stuck with it, persisted through those questions, and realized that you could reduce externalities through building economies of scale, staying ahead of regulation, creating subbrands that were willing to pay more, and asking competitors to join cross-collaboration industry associations that are raising the bar and lowering the cost for everyone.

We stuck with it because we knew the cost of those unaccounted for externalities could no longer be ignored. You have to believe that elements of capitalism are broken to want to change capitalism from the inside. And you have to love it for what it does for economic mobility to believe in it and dedicate twenty or thirty years to saying the thing that broke it has a role in fixing it.

Pagitsas: Given your long and deep experience in the sustainability field, have you seen a shift in the perception of sustainability and ESG requiring "soft skills"? Why are women are traditionally associated with the field?

Mendelsohn: I want to be very, very clear. It's not that ESG is soft. It's not that women are soft and that, therefore, these things go together. Because of the origins of ESG in philanthropy and PR, it was historically acceptable for women to touch. It was acceptable to go into this "optimistic career" that didn't own a P&L [profit and loss statement].

Traditionally, sustainability work has largely been led by junior employees, frequently women in the middle management level. If we go back to say the early 2000s, sustainability positions were positions that were safe for women to take because they were about doing "the right thing." They were about stakeholder engagement and management. They were about PR. They were about stories. From the beginning, the concept of environmental and social impact and stakeholders that engage with your business and your brand through environmental and social consequences was painted as "soft."

Sustainability officers did not own P&Ls prior to 2019 writ large. Yes, you will have leaders who own P&Ls at solar companies. You can have any number of P&L-owning folks at a recycling company. But I'm talking about my area of specialty, which is corporate affairs, and helping the board and the CEO through the transition of taking ESG from sitting on the shelf to integration into the business.

To be a sustainability leader in the prior decade meant being "overhead" or a "nice-to-have" versus being someone generating revenue. But what happens to "nice-to-have" when profit takes a downturn? It gets cut. So, it became more socially acceptable for women. Now, I have seen that change. It's a dramatic shift.

Pagitsas: How would you summarize your philosophy about the role of sustainability and business? And how have you approached your career as a result?

Mendelsohn: It has been absolute fidelity to the idea that environmental and social challenges would be solved fastest and at scale, with public and private money resources. It's been declining traditional commercial roles that offered more money. And it's been about finding a niche within ESG. If someone outside of ESG thinks ESG is niche enough, they're incorrect. It's a big space because ESG is big.

Pagitsas: At the end of the day, how would you describe your job as a CSO for a Fortune 500 company?

Mendelsohn: I just do business. I protect profitability and support revenue generation. I engage associates to attract the best talent and help my company retain them. I make sure that we are the stewards of our clients' brands when they choose to partner with us, and I make sure that our technology solves clients' environmental and social pain points.

I capture all this in data, stories, and trends, reporting it publicly to investors. At the end of the day, after going through all that, we're actually where we started, which is returning value to shareholders by creating a more livable and stable society.

Brian Tippens

Vice President, Chief Sustainability Officer, HPE; President and Executive Director, HPE Foundation

Brian Tippens is VP, chief sustainability officer (CSO) at Hewlett Packard Enterprise (HPE). HPE is a business-to-business company providing information technology (IT) solutions, including servers, storage, networking, and consulting, with over $27.8 billion in revenue through the fiscal year ending October 31, 2021. Brian oversees a global team focused on driving measurable and transparent environmental, social, and governance (ESG) strategies that benefit their customers, investors, employees, and the communities in which they work.

Brian is also president and executive director of the HPE Foundation, whose mission is to deploy skills, resources, and technology to address social challenges and advance the way people live and work.

In his prior role as VP, chief diversity officer at HPE, Brian was responsible for providing vision, management, and strategic planning in championing an innovative culture endorsing inclusion, diversity, and equity. He held previous leadership roles in HPE's global procurement and global real estate organizations. Before joining HPE in 2000, Brian was at Intel Corporation, providing legal support for manufacturing.

© Chrissa Pagitsas 2022
C. Pagitsas, *Chief Sustainability Officers At Work*,
https://doi.org/10.1007/978-1-4842-7866-6_11

He serves as a director or an advisor at several organizations, including Walker's Legacy and the National Action Council for Minorities in Engineering (NACME). Brian was recognized as one of the 100 most influential African Americans in business by Black History Organization, a top executive in corporate diversity by Black Enterprise magazine, and a top 50 diversity professional in industry by the Global Diversity List.

Chrissa Pagitsas: IT services and products are a large business sector. What are HPE's services and products within it? Who are your core customers?

Brian Tippens: HPE is a multinational firm based in Houston, Texas. Our company provides technology solutions to businesses around the world to support customers across a variety of industries, from manufacturing to agriculture, oil and gas, and other industries. While we offer hardware, software, and infrastructure, we have pivoted our entire business model to be a service-based organization that delivers all our products as a service. To put that in context, our offerings span cloud services, compute solutions, high-performance computing and AI [artificial intelligence], intelligent edge, software, and storage, which can be purchased outright or delivered via a service-model offering to the public sector and small, medium, and enterprise businesses.

Pagitsas: What is HPE's sustainability or ESG strategy? Why are you investing in it?

Tippens: Our mission is to create sustainable and equitable IT solutions that meet the technology demands of the future. To achieve our mission, we focus on leading the low-carbon transition, investing in our people, and operating with integrity. Why are we doing this? We believe our ESG strategy is critical to growing and evolving our culture and providing value to all our stakeholders. The integration of ESG issues into our business strategy increases the competitiveness and resilience of our business and differentiates us in the marketplace.

Our strategy is driving toward the low-carbon economy, which includes reaching net zero across our value chain by 2050. We, like many companies, have intermediate goals to help us reach these goals, including ones focused on renewable energy and the reduction of emissions from our operations or suppliers' emissions. We are also targeting that eighty percent of our production suppliers set science-based targets by 2025 and increase our products' energy efficiency performance by thirty times from a 2015 baseline. We're also working on our road map to 2050 that compartmentalizes the net-zero goal in a more proactive way to help us understand what investments we need to make to get there.

Our climate strategy also focuses on how we equip our customers to get to their net-zero goals, as the largest portion of our footprint is our customers' use of our products. As a provider of IT services and solutions to enterprise customers around the world, we can make a big impact by helping them reach their goals by focusing on energy efficiency in our supply chain, but also in the products that we produce.

Next, a big part of our ESG strategy is investing in people in communities globally and internally through our focus on inclusion and diversity. We've been quite public about the representation of executive women and African American, Hispanic, and Asian ethnicities globally within HPE. We've had goals around diverse representation at all levels of the company. We're always fighting to expand the measurement of diversity with other human capital metrics, such as pay parity and living wages. We also focus very deliberately within our ESG strategy on the impacts we make on communities.

We're now building a new ESG team. We took what was a very environmentally focused team and began to take a broader environmental, social, and governance view. Within my new organization, we brought in the HPE Foundation, our non-profit organization where I serve as president and executive director, to very deliberately signal that our strategy had to bring in social aspects and vehicles like the foundation to make a positive impact on communities. Through that channel, we focus on disaster recovery, disaster relief, STEM [science, technology, engineering, and mathematics] education, and several initiatives to support our broader goals as a company.

Finally, we're emphasizing ethics and compliance to make sure we're working in an ethical fashion, which is where the *G* of ESG comes in. This scope includes everything from administering our standards of business conduct training that is held throughout the year for our team members at all levels to ensuring ethical sourcing of conflict-free minerals and addressing issues of human trafficking and modern slavery in our supply chain. These are the three primary pillars of a broader ESG strategy and approach that we're refining now.

Pagitsas: Some companies only focus on environmental issues, and others have separate leaders addressing environmental and social ones. What was the internal conversation that drove the decision to bring the environmental and social issues under one organization under your leadership?

Tippens: It's been a bit of an evolution for us which mirrors, I think, the situation for every company. Our evolution was prompted by the departure of my predecessor in the CSO role. The vacancy prompted us to step back and consider what was best for the program. We asked ourselves, "Do we hire somebody with the same profile as the prior CSO—an environmental leader with an environmental science degree?" While we felt we could hire someone externally who is a specialist in this space, we felt it was an opportunity to address the external pressures we were feeling by looking internally for someone who knew our business.

Pagitsas: What are those pressures and who is exerting them?

Tippens: We're hearing more from our customers, our investors, and the analyst communities that pay a lot of attention to our transparency at all levels and on all issues, not just environmental issues. Customers alike are increasingly factoring sustainability criteria into their purchasing decisions. In fact, HPE receives over 1,300 inquiries each year for information regarding HPE's sustainability practices. We also see investor interest growing in environmental, social, and governance issues—making sustainability a priority for executives at any large company.

Pagitsas: So, in evaluating the scope and expertise of the next CSO, you and HPE leadership wanted someone who would focus on addressing the needs of those stakeholders, knew the HPE business, and would centralize disparate functions.

Tippens: Yes. We'd had a segmented approach to social issues across the enterprise and didn't have one leader with oversight. We saw that it was probably time in our evolution to have one center of expertise and excellence around all things ESG recognizing there will still be pockets that connect with the human resources, supply chain, and other teams across the enterprise. However, from a business standpoint, we knew that it was the right thing to do to bring all these pieces together.

In addition to thinking about investors and analysts, we've got a small part of our team that spends a lot of time with our customers listening to the voice of the customer and acting as a consultant on energy efficiency and digital transformation. We think about how our products and services can help them along their journey. While those conversations began around the energy efficiency of this server or that storage, they expanded over the years to discussions about what's the customer's broader sustainability story. What is the customer doing in terms of supplier diversity? What is their diversity and inclusion story?

Taking all these pieces together, the need for a cohesive environmental, social, and governance strategy focusing on investors, customers, employees, and community, we saw the business case was there to bring together the existing teams and elevate the *S* in ESG. That's where I came in. It was a good opportunity to bring me in as a leader with more experience in the business, social issues, existing knowledge of the company after being here twenty-plus years, and good visibility and exposure with the board of directors and executive committee. For the last five years, most of my time was spent on the social side of ESG as HPE's chief diversity officer. Before that, I did stints in other roles within the organization. This is my first time bringing all those disciplines together. It was a very deliberate move on our part.

Pagitsas: What was the business case for HPE to move to this broader ESG strategy?

Tippens: It was simply around leveraging sustainability and leadership to accelerate our business transformation. It's not viewed as separate by any means. It's core to our business and our transformation to pivot everything we do as a service. Whether it's meeting compliance-level requirements or the scrutiny of the Securities and Exchange Commission [SEC] in the United States and its counterpart in the United Kingdom, we never want to be in a position where we miss a sustainability lever that is a condition of doing business.

We can use our ESG leadership to support our leadership position in markets because our products are more environmentally sustainable. We can also provide consulting services to our customers, including meeting our customer's demands around a broader sustainability story. For all those reasons, ESG is helping accelerate our transformation.

Pagitsas: Do you think more companies will move to this organizational structure as they deepen their ESG and sustainability strategies?

Tippens: Yes. Last week, I spoke at the 2021 Corporate Eco Forum, which HPE is a member of. I went to that conference thinking I would be the odd man out. I would come into a room full of chief sustainability officers with climate science degrees from prestigious universities. [chuckles] That wasn't the case. I found people with lots of mixed backgrounds. I saw folks with backgrounds in chemical engineering or from the marketing side of the business that brought in different points of view who are expanding their knowledge about all those different pillars and having very strong teams.

Kara Hurst of Amazon is a great example. She's got a high level of ESG expertise but has an amazing team of specialists and engineers. As we see evolution in the space, we'll see more leaders focus on both environmental and social issues, bring in foundation work, address diversity and inclusion and tackle a variety of interrelated issues as chief purpose officers, heads of ESG, or whatever the title is. The evolution at HPE was easy for us because we had a business case that supported the fact that we see value from making this move.

Pagitsas: What is the new ESG team structure?

Tippens: Our new ESG group has a dedicated team of about fifty or so people. Within it, there is a team that looks at ESG reporting and rankings and engages with investors. Another team focuses on sustainability innovation, what's coming next, and how we design for sustainability into our supply chain and our products and services. We also have a team that focuses on sales enablement and works with our customers around the world as consultants. We are building out a team around social issues and have a team of environmental strategists to drive these topics across the portfolio. Lastly, we have the product compliance and engineering teams.

Pagitsas: It's unique to have a compliance and engineering team within an ESG team. Does this team represent the G of ESG?

Tippens: Yes. I think it's a combination of E, S, and G on some level, but it's all-new for us. I view it as a holistic end-to-end approach as we take the sustainability strategy and put it into action. Historically these teams have been separate from the sustainability teams.

Pagitsas: What does the product compliance team do?

Tippens: The team first addresses market access. Product compliance engineering thinks about whether the product's power supply meets a certain country's electromechanical and radio frequency requirements. Given our broad product set, someone has to ensure that engineering products meet regulatory requirements in every country and work with governmental agencies to understand emerging requirements because we always want to be sure we can sell products in every market.

However, this definition is the team's minimum level of competence and scope. For example, if there are governmental requirements around ozone-depleting substances, the team could tick the box and stop there. We expect that they work deliberately with the product generation units to be compliant and meet requirements ten years down the road but also make sure that we're thinking about engineering for sustainability. That's the ideal. The hypothesis is that by taking on these teams that were initially compliance-focused, we can bring in not just compliance and market access but also leadership built into our products.

Pagitsas: It sounds like you are building a dynamic organization that can address current business needs and flex to address future needs. Is that true?

Tippens: That's right, we are trying to figure out what the future will look like. You can do that in multiple ways, such as pay consultants. Nothing against any of those consultants! But generally, they don't tell me that I'm still going to be able to sell products in Kazakhstan, Angola, and everywhere else in the world where there are nuanced differences. Some of it is understanding and thinking about the future. With a presence across the globe and knowledge about emerging regulatory requirements and all those facets, we can be the team to help connect the dots across our product generation units.

There are particularly aggressive, forward-thinking geographies, such as the Nordic countries and California in the United States. Our products are designed to meet, and at times, exceed regulatory requirements as we work with regulators and government officials in various geographies to help support their agendas and leverage the resources that we have in our government affairs. Our team understands the emergent requirements but also, on some level, drives them.

Pagitsas: Clearly, you've spent a lot of time thinking about organization design at HPE. How do you think the CSO role and teams will evolve in the future?

Tippens: I think a lot about the evolution of the CSO role, whether it's called chief people officer, chief ethics officer, or head of ESG. I think of it as a "CEO minus one" role that focuses on these environmental, social, and government spaces. I think we're going to see a convergence of all these spaces, where pieces that have lived in different parts of the enterprise—the environmental team over here, the HR and the diversity inclusion teams over there—are going to come together.

Pagitsas: Turning to the IT industry, what role do technology firms have in tackling environmental issues?

Tippens: As an industry, IT firms are not typically huge emitters of greenhouse gases in comparison to the oil and gas and manufacturing industries where you want to drive the biggest reductions. Software firms, technology firms don't tend to have a huge carbon footprint, at least in their own operations. But we're in a unique position to think about our customers' use of our products and how we will assist them and help drive down their footprints.

As we're all working toward our net-zero aspirations, everyone's going to need to manage big stores of data or figure out how to operate their own IT portfolio, whether they do it on-premises, put it in the cloud, or use a hybrid approach. Technology companies like ours are positioned to help fix the problem, not just for ourselves, but to be a solution provider to everyone else in the world.

Pagitsas: What is an example of a sustainability problem that you think could be solved with IT?

Tippens: Let's take the example of a major hospital system that is managing legacy data centers it accumulated from buying up regional hospitals for their network. The system is making a big investment in each hospital's data center. It's a lot of energy use because everything needs to stay nice and cool while those servers run 24/7. That takes up a lot of resources and manpower to manage.

When a hospital is thinking about its carbon footprint, it wants to be a good citizen and set net-zero targets while managing the data centers and the needs of its customers, the patients, and doctors, nurses, and support staff. The hospital may be growing in terms of the number of patients it serves while also managing some serious privacy requirements. People might tell the hospital's administrators to just put everything in the cloud, but HIPAA [Health Insurance Portability and Accountability Act] and other requirements might mean they need to have some data stored locally or because doctors need quick access to information. HPE can come in and say, "We can help you with a hybrid approach. We can help turn everything on as a service. We can manage your data centers for you and use our platforms to help plan your network and see real-time network performance."

Working with them, we can take those legacy data centers off their hands and decide which portions of the data centers can live in the public cloud and which stay on-premises. Over time, they see that their electricity bill and their carbon footprint go down significantly because they no longer have to manage those legacy resources but can meet the needs of their customers. If we do this right, they can provide better and faster service for their customers, the patients.

Pagitsas: Would you argue that there's a social benefit that it's better patient care?

Tippens: Totally, there is better patient care. For example, with better technology, doctors and institutions have quick access to information, and there's more "healthcare portability" than there was before. Now data pools can be analyzed, and solutions created based on larger data sets. Data is the new currency. The faster we can gain insights from the data, the faster we can create solutions that can be applied to the population. COVID-19 vaccine development is an example of this.

Pagitsas: What is an example of how HPE and its clients use IT solutions to address environmental issues?

Tippens: An example is in precision agriculture. For example, we can use technology to measure how much water crops need. We have a pilot project right now with a government entity in the United Kingdom where we're looking at the effects of climate change on a particular area of crops and what a major flood would do to the area. All the data we have can help us with predictive analytics.

In the manufacturing industry, we look at the manufacturing life cycle and the very beginning of the cycle, which is at the prototyping phase. An incredible number of resources go into creating prototypes. You're using less material and resources if you can do that digitally and go through the prototyping cycles faster. When you move into the manufacturing phase, you can identify the amount of downtime if equipment or a product goes bad by using data and technology to predict some of the maintenance time or the issues coming down the line. You can save an incredible amount of time and resources.

Pagitsas: How is HPE using technology to rethink the sustainability footprint of the IT products it manufactures?

Tippens: We're rethinking what "end of life" means, that last phase in the manufacturing life cycle. We're looking at the problem of electronic waste and trying to make sure that we are not a contributor to that. We're designing our products so that they can be easily upgraded or repaired. And, instead of using the term "end of life," we say "end of use" because products can go to one of our technology renewal centers [TRCs] in either Andover, Massachusetts, or Erskine, Scotland to be refurbished and resold for secondary use.

The TRCs are the largest IT manufacturing refurbishing centers in the world. We take in IT gear regardless of the original manufacturer. We break everything down to the component level, rebuild the products, and sell them on the secondary market, so we're giving products a new life. In 2020, we were able to re-manufacture and re-market almost ninety percent of the three million products that came through the technology renewal centers, which is a pretty fantastic rate.

Pagitsas: Shifting to the S of ESG, how do you define social issues? Why do social issues matter for businesses?

Tippens: I would say that the definition of social issues is evolving because they are so complex and broad. I've been talking with other sustainability professionals about the essence of *social*, and there's no general consensus on what it means.

When you talk about the S, most people gravitate toward volunteerism, philanthropy, and community giving, and that's certainly part of it. Our HPE Foundation builds on the legacy of the HP Foundation and focuses on how we can leverage technology to solve the world's biggest problems around STEM education, for example. That's one piece of the definition of social. We also always want to measure community impact through lives impacted, dollars donated, and employee hours volunteered. That's important.

However, I think more fundamental types of social issues start with the more basic stuff that's mostly in the human resources domain around human capital metrics. Do we have a representative workforce that looks like the communities we serve in terms of gender, ethnicity, and other measurable categories? That's at the core of *social*. It goes beyond that to issues like having pay parity and a living wage. It can also be issues like the type of benefits we offer for same-sex partner relationships, maternity/paternity leave, learning and development, and all those things that help advance the lives of our team members.

We must look at the "people" impact of our operations by auditing our factories to make sure they've got fair wages and work hours. We're looking at issues of human trafficking and modern slavery in the supply chain.

Pagitsas: How do you measure the social impact of HPE's business?

Tippens: I don't think there's any consistent impact model or framework yet, but I think though we'll get there soon. Today, investors and analysts are asking for answers to questions that are easy for us, such as, "What's your diversity scorecard look like? What's your pay look like? Do you incentivize your senior leaders to meet goals around environmental sustainability?"

It'll evolve to a point where we'll have a similar holistic framework in the same way we have one for measuring a path to net zero for carbon, but I think it will take some time to evolve. At HPE, we're building a new social pillar and expect its new leader to be okay with a little bit of ambiguity. With social issues, we need to take an approach that says, "Here's the general pieces in the buckets and the things we think we want to measure, but we need somebody to come in and build it."

Pagitsas: Yes, ambiguity is the name of the game in the social space because of the vast number of potentially applicable metrics. It's also challenging because the impact may only materialize after a long duration and is alterable by many variables outside of a business' control.

You've touched on the diversity of the workforce as being a critically important social issue. What are your thoughts about representation by people of color within the sustainability field?

Tippens: There isn't a lot of diversity at the sustainability conferences. On one level, it's a pipeline issue. There is an underrepresentation of minorities that go into the fields that tend to be feeders for sustainability leadership roles within business, such as climate science and environmental engineering. While the background of the senior sustainability leaders seems to be changing, as I mentioned earlier, there is still underrepresentation in the a field's broader feeders.

Even if a sustainability leader is coming from a different background, it's a larger issue of representation at senior levels of a company. Chief sustainability officers tend to be fairly senior leaders in the company. Even when we do well in diversity and can change the numbers on minority representation overall at the company, they are not reflected within the senior leadership seats. You've probably got fewer people of color in the sustainability field and then fewer who are rising to the top of their fields. That's an issue.

We need to approach the problem deliberately—in a way that I don't think we, as an industry, have. We've done a lot of work broadly across the industry to focus on different pockets, such as providing STEM education to increase the number of African American and Hispanic coders. However, I don't think we've been as deliberate or thoughtful to prepare a diverse talent pool for the CSO role or ESG leadership as a discipline. The field is growing and evolving rapidly and is not going away anytime soon, creating an opportunity and pathway for deliberate investment.

Pagitsas: Pipeline development is critical and needed, but it is a long-term solution. Is there an immediate opportunity to increase the diversity within the sustainability field?

Tippens: While there's a shortage of diverse leaders in sustainability or ESG, there's no shortage of African American and Hispanic leaders in diversity leader roles. Perhaps the low-hanging fruit is to take distinguished leaders in the inclusion and diversity space and broaden and challenge them. If you've got the right team structure, the leader doesn't need to be the top-of-the-class educated climate scientist or know the alphabet soup of ESG ratings to be successful, right? They need to be a leader that understands the nuances of leading a major, multi-faceted program and can work collaboratively with peers inside and outside the organization. Maybe the first step is to expand the opportunities for these folks.

Pagitsas: That's a great approach that companies can apply today. Let's talk about what drives you as a leader. What excites you about your work and role as a senior executive at HPE?

Tippens: I would say a few things. One is an opportunity to really drive growth. I've been with this company for twenty-plus years. I never thought that was going to be the case. It's rare to hear that, particularly in tech, but I've been through many different iterations of HPE. It's a company that has changed a lot over the years because it's gone through a lot of acquisitions and divestitures. As the shape of the company has changed, so have my roles. I'm excited to be at a level where I can help drive that positive trajectory for our growth and make a business impact and do some good. I'm purpose-driven at the core. I'm naive enough to think I can change the world. I'm driven by a desire to do good and drive the company's business growth.

Second, as a leader, I gravitate toward roles that are business intellectually challenging. I've been in the CSO role for six months, and I think the space has probably changed more in the last six months than it has in the last six years or sixteen years. It's a great interesting space.

At this point in my career, I enjoy building a legacy behind me. Coming into this role, I took a very small flat team of environmental sustainability experts and built-in management positions and clear career paths. I also built out real infrastructure that will live on well past me. Building that infrastructure and creating the pathway for the next generation of leaders is super important to me. It's very exciting building something that's critically needed right now in the industry, and the need is not going away anytime soon. I'm not going to be the leader to see it through over the next ten years, but I am putting the right things in place for it to succeed over the long term. That's always exciting. That drives me as a leader.

Pagitsas: What was your career path to the CSO role at HPE?

Tippens: I became a lawyer because I grew up at a time when that's what you were supposed to do—go to business, medical, or law school. I started my career in technology while going to law school. After coming out of law school, I did legal support roles in the technology space for several years. Even while I had career satisfaction with doing good and meaty work around intellectual property and licensing negotiation, I would spend as much time of my time, if not more, putting on the company polo shirt to go help volunteer in communities on the weekend or do the Habitat for Humanity home build. One of the programs I enjoyed most was supporting our supplier diversity program, going to the trade shows, and talking to minority vendors who wanted to be suppliers to HP. That was exciting because it helped small businesses in a business context and created wealth in underrepresented communities. That was a bit of an "aha!" moment for me. I was much more rewarded doing that work than the legal work that was my day job.

In 2006, I jumped at the opportunity to lead our sustainable supplier diversity program and leave the legal role behind. In fact, I voluntarily took a demotion to move from this legal level to a slightly lower-level position in the supply chain to drive a program that I was passionate about. It was the best career move I ever made because I was able to go from an individual contributor to a supervisor to a manager until I had grown this supply chain program focused on entrepreneurship and small business diversity. As it built up my name as a thought leader in the space, it's probably the most distinguishing career step I've ever made.

I expanded that program, became a procurement leader, and took on some adjacencies related to environmentally sustainable procurement and procurement compliance, which allowed me to grow as a procurement leader and executive. In 2012, I realized it was time to do something new. I made a leap and went to our corporate real estate department because I knew I needed some operations on my résumé. If I was to be a big senior leader in the company, I needed to prove that I could move the big rocks.

Within the real estate department, my role had a wide range of responsibilities, ranging in activities from bathroom cleanliness to food cafeteria satisfaction. My team had to make sure the CEO always thought the bathroom was clean. Our chief information officer at the time bought a sandwich in the cafeteria headquarters two days in a row and thought the bread was stale. I had to spin up a tiger team around bread quality and find new bread vendors. While these small examples may seem inconsequential, it was a good role because it helped me understand and learn more about the operational side of the business. I also did have the environmental sustainability of our real estate footprint in my portfolio.

In 2014, the company asked me to come over and take on the role of our chief diversity officer. I pushed back at the time for reasons we discussed earlier. I didn't want to be the only African American professional at the senior leadership table. I didn't want "to diversity" myself into a career corner. I knew the work was super meaty, important, and needed, but I didn't know if I wanted to be that leader. Ultimately, I took the opportunity when the CEO said, "We need you in this role at this time." It was at a time when the lack of diversity in tech was extremely apparent and frankly companies were being scrutinized over it. We needed to navigate our way through some tough water. I took that on, and in the five years that I had that role, it felt like we could do some good things from diversifying our board, putting lots of good policies in place, increasing diversity within the company, and focusing on inclusion. It was a great five-year ride that essentially led me to this role as CSO, which explores the adjacencies of what I'd already done and allows me to operate in a new space.

Pagitsas: One theme in your career is proactively moving yourself to new areas of the company to gain new skills and expertise. Another theme seems to be that you influenced critical decisions for HPE's success, such as supply chain diversity and the operation of its facilities, without owning the profit and loss statement [P&L] of a business line or product. This combination of responsibility and accountability is a frequent situation for CSOs.

What are the enablers of success for a CSO at a billion-dollar multinational company but who does not control a P&L?

Tippens: Using my own experience, there are several factors. First is the background and makeup of the leaders. When I came into the CSO role, I already had twenty years in the company, during which I interfaced with senior leadership and built relationships with them and the board. I came in with a high level of credibility. That's important as companies think about leaders that they need to put in these CSO roles. It needs to be somebody who's got that level of business credibility who can justify significant new cash investments to expand teams and invest in areas.

Another factor is structural. You need to build a structure that supports consensus, cooperation, and influence across the company. At HPE, we have our Living Progress Strategy Council. It's a group that convenes regularly, and its representatives are from across the business. It's an example of how to embed sustainability into the governance of the company.

The most important factor is having those in-person interactions, that rapport, to show the business leaders with P&L responsibilities that sustainability is not just a corporate mandate. That's the key. I'm here to help make the company successful. I'm also here to help make each of those leaders successful. I can help if they have compensation tied to performance in this space, or I can help them with their sales pitch and markets because my team has a unique view as to what customers are saying in this space.

CSOs need to come in a very collaborative, cooperative way and not just be perceived as implementing "yet another corporate mandate," which has historically been the downfall of diversity and inclusion teams. They have set aggressive goals in the silo of the human resources group and then slap the hands of leaders who don't hit them.

With sustainability, we can't position this as "extra." This has to be positioned as core to the business in a credible way—not just ticking a box on compliance but being a differentiator in the market. Our goal is to interact collaboratively with leaders using a carrot, not a stick.

Pagitsas: Is there a story that exemplifies your teams' collaboration with a business leader to create business value and simultaneously create environmental or social impact?

Tippens: I'd say the most recent was on the S side. We had a recent leader take over our networking business acquisition. The business's founder recently moved on after leading it for a decade. The new leader and I set time aside and talked. He was so new to the business that he didn't know a lot about the bright spots that had taken place in the past. We reminded him, "Your team did some amazing things during COVID-19 where we took your Wi-Fi connectivity kits and helped wire up schools and hospitals."

During the height of COVID-19, you may have heard stories of students who didn't have connectivity at home. Sometimes the best thing they could do was get in the car and drive to the school so that they could do their lessons in the car. There was a lot of HPE technology used to build kits to solve that problem. We did that reactively for COVID-19, but we sat down with the new leader and discussed continuing that program in inner-city schools and hospitals. It's not all done yet, but we're going to make something happen there.

That was a bit of a collaborative "aha!" where we could come together and create a winner from both social and business standpoints. The business leader knows that's going to distinguish him in the market against competitors with the good work that we're doing. That's an example of collaboration between sustainability and business teams that stands out.

Pagitsas: That seems is a great example of integration of ESG principles into a core product. What tensions or challenges do you foresee as sustainability becomes more integrated and critical to businesses' operations and success? And lastly, where are you optimistic about the future?

Tippens: At some point, I think we all need to decide what we stand for as companies. I think about a Venn diagram capturing the strategy work around ESG. One circle on the right side is driven by investors, the SEC, the ratings and ranking agencies, CDP, and the Dow Jones Sustainability Index [DJSI]. It's the ticking on the box on everything we have to do. Over on the left, there's another circle for making the world a better place, driving toward creating impact, being a force for good, and taking a position on social issues.

If we are to change the world we live in, there's this overlapping of the circles, but I fear the focus on the right while critically important can hinder the focus on the left. Why? If we get to the point where we've got to get sign-off from our legal teams because these measurements will be reported in 10-Ks, we're not going to make any bold positions. Some of these metrics are still evolving, as I mentioned earlier. It may stifle the success of an ESG strategy if we're not deliberate about how we build out our teams and our approach to handling both circles.

To change the world, we must remain focused on being a force for good and driving meaningful and impactful change. Incremental efforts won't be enough. We need transformative change that requires collaboration between public and private sectors, and I believe technology will enable a more sustainable future.

James Gowen

Senior Vice President, Global Supply Chain Operations and Chief Sustainability Officer Verizon

---

James "Jim" Gowen is senior vice president of Global Supply Chain Operations and chief sustainability officer at Verizon. Jim is deeply involved in advancing innovative and sustainable technologies and spearheaded the issuance of Verizon's first $1 billion green bond. Through Jim's leadership, Verizon has issued $2 billion in green bonds as of 2020 to provide capital for Verizon's renewable energy, energy efficiency, water conservation, and reforestation projects. Jim has implemented a carbon intensity metric and paper suppression, waste reduction, recycling, and management of end-of-life material recovery programs.

In his supply chain role, Jim leads all inventory planning and logistics operations in the 150 countries in which Verizon operates and has responsibility for Verizon's Global Manufacturing Operations.

Jim is a member of Penn State University's Smeal Sustainability Advisory Board and the current chairman of the Global Enabling Sustainability Initiative (GeSI).

Chrissa Pagitsas: Verizon is ranked twentieth on the Fortune 500 list with over $128 billion in revenues in 2020. Tell me about the breadth of Verizon's business services and operations that drive this rank and revenue level.

© Chrissa Pagitsas 2022
C. Pagitsas, *Chief Sustainability Officers At Work*,
https://doi.org/10.1007/978-1-4842-7866-6_12

Jim Gowen: Verizon operates in 150 countries providing comprehensive consumer, small business, and enterprise services for both wireless and wireline technology. Our multi-use network at scale provides our customers with the services and solutions they need to stay connected and thrive in the digital economy.

Our consumer and small business services are primarily focused in the United States, where we are the nation's largest wireless provider. This part of the business excels in three key components—home, value, and mobile. As we have seen from the pandemic and the transition to remote work and school, customers rely on the network and stay for the unique experiences we offer that are most relevant to their lives, whether they are at home, school, work, or on the move.

Verizon supports almost every Fortune 500 company in the world at various levels. The demand for 5G technology is growing as enterprise customers look to enable solutions like computer vision, augmented reality, and machine learning. For large corporations to address their sustainability and responsible business goals, these solutions are essential as they are designed to increase productivity, provide enhanced security, and reduce latency.

Pagitsas: Given the diversity and breadth of these businesses, what is the scope of Verizon's sustainability strategy?

Gowen: We believe in "Work green. Live green." We build sustainability into how we operate our network, from our 130,000-plus employees around the world to our suppliers to the devices in our customers' hands, helping them reduce their carbon footprint. Our customers want to know that network coverage will be there for them when they pick up the phone, turn on the television, laptop, or gaming console—at home or on a mountaintop.

As a responsible business, Verizon knows that the network must be as efficient, robust, and resilient as possible. Our network is who we are at our core. Running a sustainable network via a sustainable supply chain is in our DNA.

As we look across the supply chain, we are putting CSR [corporate social responsibility] standards in place to ensure that our vendors are thinking about the triple bottom line alongside us. For example, we make sure that our large radio vendors build their radios as sustainably as possible. In other words, we are asking about the kind of parts they use and how they are designed with end-of-life in mind, and if they are creating products for a circular economy when it comes to 3G, 4G, and 5G.

Verizon is working very hard to integrate sustainability into every experience for consumers. When they go into a Verizon store or building, they see the US Environmental Protection Agency ENERGY STAR sticker certification on the front door. Or perhaps they will see solar panels in the parking lot of our offices where we have installed renewable energy sources on site. Verizon is in its eighth year of winning the ENERGY STAR Partner of the Year Award, and we have a goal to achieve net-zero operational emissions by 2035. We are very proud of that.

When they are inside the store, customers will see options for cases and accessories made with recycled content and opportunities to turn in old devices. We have a significant and far-reaching reuse program as part of our circular economy for devices. When a customer is ready to upgrade to a new device or get another piece of equipment, like an earbud, they are incentivized to return their old device so that we can give it a second life. These are just some of the ways Verizon is helping customers be more environmentally responsible.

Pagitsas: How do you make this vast supply chain also resilient?

Gowen: It's about asking the right questions and knowing the source of the materials and labor for the products that make up your supply chain. It is imperative that you have the right partners in place who can not only meet our high standards but who are transparent in their own operations and can easily answer. Where are you manufacturing? How are you manufacturing? Where does your core material come from? What redundancies do you have in place? Is your supply chain diverse and flexible enough to manage through unexpected events such as natural disasters, political unrest, and economic turmoil?

A sustainable supply chain must care for the geopolitical and macroeconomics while also being 100 percent cognizant of the environmental aspects of a changing climate that is increasing more every day.

Pagitsas: Sustainability strategies are incredibly interdisciplinary. Where should that team sit within the organization for the strategy to be most successful?

Gowen: I think you need to place the team somewhere in your company's core operations to ensure it is not siloed in corporate functions. Embedding sustainability within operations ensures a different level of visibility, accountability, and commitment within the organization. This is what I've seen done at peer companies that have been the most successful.

Pagitsas: Let's talk about how you partner with your C-suite peers and the board. You can't do it alone, right?

Gowen: If you look at the last three years, there has been no better champion than Hans Vestberg, our chairman and CEO. He lives, breathes, and understands corporate social responsibility and sustainability. He has been a tireless leader in this field, and I even have trouble keeping up with him.

Hans sits on the board of the United Nations Foundation and was involved in the creation of the UN Sustainable Development Goals [SDGs]. He understands the process and the importance of setting ambitious goals because of what is at stake and the opportunity to innovate to move the business forward and all of our stakeholders, from customers to employees. From the start, Hans has been an advocate and encouraged my team and me to set lofty goals and push harder. This commitment has permeated our company, and I present to his direct reports on what is going on every other month. Each leadership team member brings their own unique views, experiences, and support to our ESG [environmental, social, governance] priorities. While they all come from diverse backgrounds and experiences, they share a desire to always push the envelope. Whether that's looking for opportunities to advance education, women in STEM [science, technology, engineering, and mathematics], solutions for our customers, finding efficiencies within the business, or reducing our carbon footprint, I have found support from each member to find ways to leverage our technology, operations, and employees to create a sustainable future.

Pagitsas: How does your sustainability strategy intersect with ESG?

Gowen: At Verizon, we believe building a better future involves making climate awareness "business as usual" throughout our organization. From investments in clean energy to upgrades to our facilities and infrastructure to reforestation projects. Beth Sasfai, Verizon's chief ESG officer, is integral to that mission. She oversees Verizon's efforts to deliver on its ESG commitments by monitoring goal-setting and developing appropriate reporting to share progress with shareholders.

Pagitsas: The C-suite is a critical partner but not the only one. What role do Verizon employees play in the company's sustainability strategy?

Gowen: You're only as good as that last person on the bench, whether that person gets in the game or not. This is something I learned playing sports throughout college and coaching ever since. This is how I approach my work. You have to value the uniqueness of each member, and we know that innovation is sparked by diversity, which speaks to the company's DEI [diversity, equity, inclusion] policy and practices. Luckily, we have a strong bench of players. Each of our employees plays a role in our efforts.

Pagitsas: Isn't a seven-person team small to run a sustainability strategy for a company with 132,000 employees?

Gowen: I always get that question. In a large company like Verizon, it would be easy to create a thirty-, forty-, or fifty-person team. However, I believe in keeping the core sustainability team small so that the responsibility is shared and everyone in the company feels ownership. This way, all of our employees understand they have a role in adopting and implementing our sustainability strategy, not just my team or me. At the same time, our seven-person team is leading some of the biggest initiatives, whether it is renewable energy, recycling, or circular economy. We are constantly being challenged to go out into the organization and build out programs locally.

Pagitsas: So, how do you build sustainability locally with employees located across the globe?

Gowen: A great deal of our motivation to do the right thing for the planet and climate protection comes from our employees themselves. We established Green Teams ten years ago through the suggestion of one employee. Today we have more than 50,000 employees in fifty-one countries on Green Teams, which mobilize employees around issues that impact their local communities and unlock their potential to give back.

For example, the Green Team in the United States runs electronics recycling rallies within local communities. We run about twenty-seven of them a year. Employees volunteer, accept your e-waste and give you a certificate of destruction. In 2020, we held only a few in-person events due to COVID-19. The nineteen recycling events we held collected more than 650,000 pounds of e-waste [electronic waste]. Verizon has collected more than 4.8 million pounds of e-waste since 2009, toward our goal to collect and recycle five million pounds of e-waste by 2022.

In Chennai, India, we have a Green Team champion there, and we have seen the team activate ideas that extend beyond environmental sustainability. For example, while the team did a tree-planting event at the facility, they partnered with Take Your Daughter to Work Day to encourage gender empowerment, interest in STEM, and women in the workforce. As you can imagine, it's a big deal, and we encourage our Green Team leaders to think outside of the box when it comes to sustainability. In Ireland, the Dublin team hosts an annual cleanup of the River Liffey, which runs through the city. It is all done voluntarily by Green Team members, and we have many more examples of how local employees have mobilized through Green Teams to bring our motto to life, "Work green. Live green."

Pagitsas: What drives you daily, Jim? Why do you get up and do it?

Gowen: First and foremost is family. I come from a large family, and I'm very blessed to have a large family. So that is 100 percent what drives me every day. Secondly, I made a very conscious decision to work for Verizon and stay with them for the last twenty-plus years. It had nothing to do with the fact that my grandmother was a phone operator way back in the day! That wasn't it at all.

I'll never forget the conversation I had with my father, who encouraged me to work for the phone company right out of college because it was "consistent." I did not buy into the idea right away, but then I thought about the life-saving services Verizon provides. When I started, there were no cell phones. It was all about 911. Fast-forward to today. It's all about first responders, the firefighters, emergency workers, and police who use our networks and devices to go into harm's way to help others. That's my passion for why I'm here and why I work at Verizon.

Pagitsas: How does Verizon show its purpose?

Gowen: Verizon has gone through an evolution in its purpose. We understand what our technology can do. We want to ensure that our resources, technology, and employees are working toward a common mission of moving the world forward through action. To that end, Verizon launched a new responsible business plan, called Citizen Verizon, to drive economic, environmental, and social advancement.

With Citizen Verizon, we adopted three key pillars that focus our responsible business goals—digital inclusion, climate protection, and human prosperity. Verizon has pledged to provide ten million youths with the digital skills training necessary to thrive in a modern economy and provide one million small businesses with resources to thrive in the digital economy by 2030. We are committed to making our operations carbon neutral by 2035 and have set approved science-based emission reduction targets. To meet these goals, we are investing in renewable energy and energy efficiency initiatives. And lastly, we are preparing 500,000 individuals for jobs of the future by enabling opportunities through upskilling and reskilling. We also recently announced that our responsible business investment to help vulnerable communities and bridge the digital divide is on track to exceed $3 billion by 2025.

We want to show up as a citizen of the world and marshal our resources to make an impact in small towns, big cities, and rural communities worldwide, with a special focus on those that are vulnerable and under-resourced. When you asked me about what gets me out of bed, yes, it's my family, but it's also what I do. I have the best job in the company, but it's so important to me who I work for.

Pagitsas: Are you saying that your company has a purpose beyond the shareholder?

Gowen: Absolutely. We look to bring shared value to our four stakeholders— our customers, employees, shareholders, and society.

Pagitsas: Are shareholders expecting anything different from Verizon these days?

Gowen: The biggest thing I've seen in the last five years is that shareholders holding Verizon stock are now asking questions they have never asked before. There is a growing interest in ESG—the environmental, social, and governance factors of the business. They want to make sure we have responsible business practices and that we are a purpose-driven company.

Pagitsas: Verizon issued $2 billion in green bonds between 2019 and 2020. Were the green bond issuances a signal to Verizon's market participation in this responsible investing wave?

Gowen: One hundred percent. More importantly, we were very deliberate and transparent in identifying how we would use the green bonds in advance and report on them annually. The green bonds are funding renewable energy, energy efficiency, green buildings, sustainable water management, and biodiversity and conservation projects. Beyond issuing the green bonds themselves, the reporting on their use is incredibly important because some ESG investors are interested in understanding the environmental impacts of our investments and how we are meeting our sustainability goals.

Pagitsas: A CEO at a Fortune 500 company asks you for advice on starting a sustainability strategy and team at their company. What would you tell the CEO?

Gowen: If I were sitting next to a CEO of a Fortune 500 company who wanted to start a sustainability practice, I would likely tell them what attributes they should be looking for in a sustainability leader. You need someone who can look not only internally but externally too. You need someone who can challenge you. When you embark on a sustainability journey, it is about a willingness and openness to say, "I don't have all the answers, and I am going to listen and learn." Next, I would tell them that when you commit to a sustainability goal, as CEO, you have to be ready to see it through and test your fortitude. And you are going to need a sustainability leader who will stick with you.

Pagitsas: Why is fortitude so important for a sustainability leader to possess?

Gowen: The COVID-19 pandemic is a perfect example. It could have been easy to put sustainability on the back burner and focus on the business. Verizon did not do that, and we kept plowing ahead with our very large sustainability goals and our day-to-day activities.

Pagitsas: What else would you tell them?

Gowen: Two things. To admit failure and to partner with others.

A perfect example is our project to convert our vehicles—like our fleet of trucks—from gas to natural gas and take it to scale. The natural gas-powered vehicles worked great. Maintenance costs were lower, but there were a lot of people who still had a great deal of anxiety around the change, and we could not get the big OEMs [original equipment manufacturers] to keep building them. This ultimately affected our roll-out, and we could not take the program to scale.

However, I had an opportunity to speak with President Barack Obama's Secretary of Transportation Ray LaHood and Secretary of Energy Steven Chu. We discussed the need for infrastructure and the role of public-private partnerships to do things better and learn from each other. It solidified my belief in the value of partnerships to find sustainable solutions. It is a critical part of my work as a sustainability leader of a global Fortune 500 company.

Kevin Hagen

Vice President, Environment, Social & Governance Strategy
Iron Mountain

Kevin Hagen is vice president of environment, social, and governance (ESG) strategy at Iron Mountain, a storage and information management services company with over $4.2 billion in revenues operating more than 1,450 facilities in over 60 countries as of 2020. Kevin advises global company leaders and business units to develop and implement sustainable business strategies that address a wide range of ESG issues, such as climate and energy footprint, human rights, social impacts, circular economy, ESG reporting, corporate philanthropy, and community engagement. Kevin leads the company's strategy to achieve its commitment to The Climate Pledge to become net zero by 2040.

Before joining Iron Mountain in 2014, Kevin led sustainable business strategy at Recreational Equipment, Inc. (REI). He also led engineering, marketing, and business development functions for companies in the renewable energy, aerospace, and military electronics markets.

Kevin serves on the board of the Clean Energy Buyers Association (CEBA), formerly the Renewable Energy Buyers Alliance (REBA), and other sustainability advisory councils and boards, such as the National Association of Real Estate Investment

© Chrissa Pagitsas 2022
C. Pagitsas, *Chief Sustainability Officers At Work,*
https://doi.org/10.1007/978-1-4842-7866-6_13

Trusts (Nareit), Sustainable Brands, and Net Impact. He is also an instructor in ESG reporting at the Harvard Extension School's Sustainability Graduate Program.

Chrissa Pagitsas: Tell me, Kevin, about your career and the path that took you to leading ESG at Iron Mountain today.

Kevin Hagen: Leaving university with an undergraduate background in engineering, I joined an aerospace and defense contractor and had the opportunity to be involved in the design and development of custom components for civilian and military aircraft and systems. It was a great opportunity because I ended up straddling the conversation between customer needs and engineering design which included the process to develop, test, and launch new products.

That early professional experience taught me how to speak multiple business languages, the customer language, and our internal design and development language. Learning how to bridge organizational silos is an important competency in the pantheon of sustainability business skills and competencies. Later I ended up based in France and really had to learn two languages.

After working in the defense and aerospace field, I joined a small startup in the renewable energy space. It wasn't much bigger than a two-bay garage operation at the time, but there was an amazing team of talent, best-in-class products, and lots of growth potential. It was very exciting because it was early in the growth of the renewable energy, solar, and wind industry, and the focus was on moving from just off-grid solutions to the grid-connected renewable energy industry. Trace Engineering was the first company in the United States to develop and deliver inverters and power electronics that could connect solar panels to the grid and power homes from both. It was a wonderful success in many ways, and we ended up going public.

However, I think focusing on a successful IPO [initial public offering] distracted us from our focus on making a difference with renewable energy. We started thinking about making money more than about making a difference and, in some ways, lost sight of the mission. I was as guilty as anyone else. As a result, the business didn't go as well as it could have gone, and I left the company somewhat disillusioned.

In 2003, I started my own consulting firm in renewable energy, and shortly after, I attended a "go green" conference in Seattle. I saw a panel with Jim Hartzfeld and Gifford Pinchot III. Jim was the sustainable business leader and right-hand person to Ray Anderson, CEO of Interface, Inc. He talked about Ray's "spear in the chest moment" that changed his perspective on business and sustainability and moved Ray to set a new sustainable course for Interface and its carpet and flooring products. Ray's story is captured in his book *Confessions of a Radical Industrialist: Profits, People, Purpose—Doing Business by Respecting the Earth* [by Ray C. Anderson and Robin White (St. Martin's Press, 2009)]. Gifford spoke not only about the vision of sustainable business but the need to train people with business experience in sustainability so they could help change companies from the inside. He, along with his wife Libba and others, cofounded a new graduate school.

Hearing the Interface story and observing what was going on with the sustainable business movement, I found myself about three weeks later on a remote island in British Columbia sitting next to Gifford at the kick-off retreat for the first MBA in sustainable business at the Bainbridge Graduate Institute. Over two years, I learned a ton. And I had an amazing opportunity to be exposed to the people working at the intersection of business and sustainability, from MIT Professor John Ehrenfeld, who taught about industrial ecology, to Ray Anderson himself, who talked about his conversion to sustainable business, and to Amory Lovins, Janine Benyus, and Paul Hawken who focused on the imperative to address climate change, and others.

Coming out of the program, I understood the literature and the thinking around sustainability as a discipline, or so I thought. I soon had the amazing opportunity to join REI, the outdoor industry retailer and consumer co-op, working to implement the sustainable business vision of Sally Jewell, who had recently become CEO.

Around this time, in 2004 and 2005, a lot of businesses had begun to experiment with sustainability. There were well-known examples like Patagonia, Timberland, and Ben & Jerry's. What was interesting about joining the co-op at the time was that it was already a billion-dollar business, and there were not many examples of companies at that scale working to incorporate sustainable business thinking. We started with the basics of metrics, data, transparency, and accountability. It led to a shift in thinking at the co-op.

In early 2013, Sally Jewell was nominated by President Obama to join his cabinet as the secretary of the interior. It was surreal to see my boss on TV at a Senate confirmation hearing. I thought that it was a wonderful mark of success for the co-op when the president introduced her as a leader in a new way of doing business. And we had the results to prove it. REI's sales had doubled from $1 billion to just over $2 billion while its absolute energy consumption went down. We had installed the first solar panels on stores and were building one of the first net-zero energy distribution centers. On the social side, we started to find ways to use the power of the co-op to advance more inclusiveness in the outdoors and the outdoor industry, trying to make human-powered sports more welcoming to everyone. The co-op was using its power and influence in the industry to lead changes in human rights and environmental performance in the supply chain, helping to launch the Sustainable Apparel Coalition, among other industry efforts.

After over seven years at REI and a new CEO arriving, I felt that perhaps it was time for me to step aside, "get out of the way," for the business to do more. So, I headed out to what I called a "ski-batical." Then a marvelous person named Samantha Joseph found me. She was leading an effort at Iron Mountain to implement a sustainable business strategy and told me that they were mountaineers who shred, and I said, "These are my people."

Pagitsas: [laughs] That's pretty funny!

Hagen: Thanks for chuckling! The inside joke, of course, is that Iron Mountain employees call ourselves "mountaineers." And "shred" wasn't slang for skiing powder. It's a reference to our secure paper destruction business. I'm not saying there was false advertising involved, but I was expecting a little more skiing!

But the opportunity was amazing. Samantha had done a tremendous job as a catalyst and as an entrepreneurial individual leading the effort to expand the idea of sustainable business at Iron Mountain. I had a wonderful interview with William Meaney, the CEO, during my hiring process. I'm paraphrasing the conversation, of course, but I heard him say that the company had a great ethos around environmental and social responsibility because the crux of Iron Mountain's brand and business is trust. The core business was to manage and protect our customers' most valuable and most important information and assets. They expected us to do the right thing, but what he was really getting at was that "doing the right thing" is not a business plan. We need to rethink business as usual. And that's what brought me to Iron Mountain.

Pagitsas: What business services does Iron Mountain provide?

Hagen: For those who may not be familiar with Iron Mountain, we are a publicly traded real estate investment trust [REIT] serving primarily business-to-business customers with over 1,450 facilities in over sixty countries. We have about 25,000 employees worldwide and annual sales of about $4.2 billion. Historically we have been in the document and asset storage business. But we realized that as our customer's information shifted from paper to digital, our business needed to change.

Today we still have a growing information storage business serving over ninety-five percent of the Fortune 1000, and we offer a full suite of data and information management and privacy solutions. From digitizing documents to becoming one of the top ten colocation data center providers in the world to becoming a leader in IT asset management, we're helping our customers make the digital transformation.

Pagitsas: What did Iron Mountain's sustainability strategy look like at the start, and where is it today?

Hagen: We've come a long way from our first sustainability report in 2014. In 2021, we announced twenty ambitious ESG goals. We have six long-term aspirational goals, such as signing The Climate Pledge, a commitment to net-zero carbon emissions by 2040. In addition, we made fourteen "on our way there" commitments, including even faster near-term greenhouse gas [GHG] emissions reductions and having women make up forty percent of our worldwide leadership team by 2025. Our goals cover all our most material ESG issues with commitments to employees, our communities, and our customers.

Going forward, a key part of our growth strategy is to incorporate sustainable business thinking—from the climate impact of data centers to the electronic waste issues associated with electronic assets to ethics and anticorruption to inclusion and diversity to human rights in the supply chain. The list of *E*, *S* and *G* issues we face is long. We know that our stakeholders expect corporations to be part of the solution, not part of the problem. We think that solving ESG challenges in our business is not only good for business. It opens the door to achieving even more as we help our customers and communities do the same.

Pagitsas: Given that Iron Mountain's business is centered on thinking about the long-term storage of information and data, how does Iron Mountain think about the longer-term horizon of sustainability strategies and goals?

Hagen: As a publicly traded company, we can't avoid measuring performance quarter-to-quarter, but we do have an inherent potential asset of being able to think a little more long-term than some organizations. Our customer relationships are measured in decades, and the average box of documents stays with us for fifteen years. As a result, I think we have something going for us. On the other hand, we don't have many of the things that some folks associate with the typical sustainability conversation. We're primarily a business-to-business company, not a business-to-consumer company. We're a service-based business, not a product-based business. Just like when I was at REI, and people said, "Yes, that can be done at Patagonia, but it could never work here." Now that it works at companies like REI, folks might say, "Well, it works at REI, but it could never work at Iron Mountain."

Pagitsas: Why do you think people initially thought Iron Mountain couldn't tackle sustainability as REI did?

Hagen: I suspect that folks who think in conventional trade-offs, either you do the right thing, or you do the business thing, look for ways to rationalize that point of view even when they see a business successfully delivering more of both. They might say, "Sure, it works at a consumer co-op, but it couldn't work at a publicly traded company."

It's a discussion we've been having at Iron Mountain over the last five or six years as we've tackled environmental, social, and governance challenges in ways that delivered better outcomes. We found ourselves challenging the paradigm that you must make trade-offs. When we figured out how to make it cost less in the first place, it got easy to advertise our long-term thinking.

Pagitsas: Do you accept or reject the idea that you must make trade-offs when businesses integrate sustainability into their businesses?

Hagen: It's conventional wisdom, which is that there are trade-offs with sustainability. It's either do the "right thing," or do the "wrong thing," or in other words, the "business thing." This trade-off mentality permeates popular thinking that "the green product always costs more," or "it doesn't work as well," or both.

Conventional wisdom says that if you really want to make a difference in your career, you join the Peace Corps. You don't join a corporation. You either do the right thing, or you do the business thing. It's easy to fall into this Andrew Carnegie mentality of making tons of money and then giving it away when it's over. It is an arguable approach given that in the early twentieth century, the Carnegie Foundation funded some fifty percent of the libraries in the United States. Yet, it's probable that we've still not come to the end of the environmental damage that the Carnegie business did in the first place.

The opportunity for us sustainability leaders in those businesses who have been given the privilege to tackle sustainable business challenges is to hold on to the tension of not accepting trade-offs until you've figured out the right business solution. That's our obligation. The reason we're going to figure it out isn't just because we're Pollyanna. It is because we have developed tools, techniques, competencies, and hard business skills that give us better opportunities to solve problems in new, better ways not to accept the trade-off.

We're more collaborative. We're more innovative. We have better data. We understand the business challenges through the lens of social performance as a business challenge. I think that gives us more tools to solve the problem in different ways. It's a symptom of the systems thinking approach. I appreciate a quote credited to President Dwight Eisenhower, "Whenever I run into a problem I can't solve, I always make it bigger. I can never solve it by trying to make it smaller, but if I make it big enough, I can begin to see the outlines of a solution."

When we figure out a better business solution, then we can do a whole lot of it. As I joke with our CFO, once we figure out how to make money at it, nobody stops us. On top of that, not only can we do a lot more of it, but other businesses can follow suit. Because now you don't have to have religion. All you do is see the better solution, then you execute.

Pagitsas: What is an example of something that now costs less, and you rejected the trade-off?

Hagen: When we began thinking about the sustainability strategy for Iron Mountain, we had over 1,400 facilities in more than fifty countries to store information, data, and documents. We knew we paid over $40 million in utility bills, but we didn't know how much electricity we actually purchased with that money. That's not because we were dumb. It's because, like many businesspeople, we just assumed that there wasn't much you could do about it that would be a big enough difference to any individual building. But in order to make our first GHG emissions footprint, we needed to figure out how many kilowatt-hours we used. That data led us to some interesting findings. For example, our electricity use was our biggest source of GHG and that there was a lot more cost volatility than we knew about. We also learned there was a lot of variability in the energy use per building, even though we assumed it was consistent. It was eye-opening when we realized how much money we could save if the worst half of our buildings had the same energy use per square foot as the best half of our buildings. The data shined a bright light into some dark places in our knowledge.

Once we realized that we had lots of opportunities to reduce energy consumption, especially in lighting, the facilities management team did an amazing job of working with vendors to develop LED [light-emitting diode] fixtures that could go in easily and cost-effectively. We thought that we would be clever and set an impressive goal to halve our electricity consumption. It was going to be great.

While we were setting this goal to cut our energy consumption, another part of the business was recognizing that our documents management and storage industry was changing. Our company had to move from being a purely document-based company or, more specifically, a physical assets-based company to becoming a much more digitally oriented business with a focus on digital storage and data center colocation. While we were promising to cut energy consumption in half, the business strategy team was promising to grow our data center business—one of the most energy-intensive industries short of mining. We were never going "to LED" our way out of that.

Pagitsas: What was your takeaway from this experience?

Hagen: This is a critically important lesson for any sustainability leader—know what's going on in your business. As soon as you become too centered on the one thing you thought you had, there's somebody else working against you without even knowing it. Silos within a business tend to be a big disadvantage and create blind spots.

When we realized that energy efficiency alone wasn't going to get us where we needed to go, we pivoted quickly to add a renewable energy strategy. If we were going to use a lot more electricity, it had better be from renewable sources because that's the only way we were going to lower carbon emissions while increasing our electricity consumption. That posed a big challenge because we assumed renewable energy was more expensive. We couldn't give our new emerging digital business a disadvantage in its cost of electricity, which wasn't going to work. But we couldn't build a new business with a fundamental ESG disadvantage either.

We could have bought a little green power at that point at a premium and called it good enough, but we didn't. We held the higher bar, in my opinion. It took us at least eighteen months with plenty of starts and stops along the way to figure out a better solution. We eventually recognized that renewable energy had a huge advantage over fossil fuel generators because they have a great handle on their long-term costs. The costs of renewable energy systems are in the upfront capital to build them. There's no recurring cost of fuel. Since there's no fuel cost, renewable energy providers can more easily have a long-term contract for a fixed price whereas a fossil fuel provider has a hard time doing that. In addition, before the wind or solar asset is built, having a long-term contract in hand for the output makes financing the project a lot easier. So, both the buyer and the seller have an interest in crafting a long-term deal.

As we brought in the finance, tax, treasury, and legal teams, and all the other folks that were going to need to financially engineer this, we were also negotiating hard with third-party suppliers for renewable energy solutions. We ended up with answers that provided not only the better, more stable long-term cost solution but also a cheaper first cost.

We saved money in the short term and paid less for green power while we locked in long-term rates. Today, we're now sourcing over eighty percent of our total electricity portfolio demand worldwide from renewable energy sources, and I don't believe we've ever paid more for green power than for fossil fuel. In sum, we went from blind spots around electricity usage and false starts around making the wrong goal setting to making real innovation happen and figuring out how to use renewables to reduce our carbon footprint. While we've almost tripled our corporate electricity consumption, we have reduced our absolute carbon footprint by more than sixty percent since 2016 while growing the business.

Pagitsas: You pointed out a lesson for sustainability leaders—know what's going on across the business. What other key lessons would you point out from this experience?

Hagen: Reject the trade-off, bring more people to the game, and get better data. In our situation, rejecting the trade-off meant hitting the pause button to say, "Wait, something's wrong here. If it costs more to do the right thing, we can't do much of it." We first had to reject that convention. Next, we had to bring more people to the table, like the finance and tax teams, to bang out the right solution for the company with renewable energy.

Last, when you have better data, you make better decisions. When we started tackling energy efficiency, we had no idea how much electricity would be consumed with the new digital strategy. We knew from financial statements how much we paid in the utility line item, but we didn't know we were rarely "on budget" for the year or the cost was going up every year. We didn't know whether that was because we used more electricity or because we paid more for it. It just was seen as the cost of doing business. This meant that we didn't have a good handle on our starting point with electricity. However, today, we know how much energy we are using, how much we are projected to use, and what we pay for it.

Pagitsas: Getting better data is an interesting point. My observation is that as you and the Iron Mountain team moved to integrate sustainability further into your services, you had to get data that you may not have looked for previously or even had. Will other companies face the same challenge?

Hagen: Absolutely. Yes, they will need to get new or better data. I think another fallacy in conventional thinking is to call environmental and social data non-financial metrics. For example, we used to think that employee diversity numbers and greenhouse gas emissions were somehow "non-financial." While true in a conventional sense, today we think of ESG data as business data. It may seem like a stretch at first, but these days if we ask the organization to help collect some new ESG data points, I commit to peer leaders in the business that if after two years they don't see the business value and use that data stream to inform their decisions, my team will stop collecting it.

Pagitsas: Why is bringing this data into business decisions challenging?

Hagen: I think that historically there has been resistance to bringing "emotional" or "heart" data, such as diversity or social issues, into business or "head" decisions. This resistance occurs because when we open the door to that discussion and start to look at some of that data, usually it's bad news. Hiding in our blind spots are often negative environmental or social impacts that we would rather not know about. And once you start measuring, it always seems to get worse before it gets better. The natural human reaction is one of two places—denial or despair. Neither is a very helpful place. You have to get to hope.

The key opportunity for sustainability change agents like us is to use that data in a way that helps people see the opportunity to do something different and get better outcomes, better environmental, better social, better business outcomes. That's the crux. We've experienced that over the last five or six years at Iron Mountain, where we went from blind spots to maybe some denial or despair and arrived at hope. Because we saw this different way to do it and now, we celebrate a better business outcome. That gave us incredible power to look across the business and see more opportunity. It was a formula, not a fluke. We can apply the formula to many more things across the business and get better outcomes.

The next stage has been to rethink even that. So far, we have challenged ourselves to reduce our negative impact, to have fewer safety incidences, to have a lower carbon footprint. There is a point at which it gets unsatisfying to make these incremental differences. In fact, I would argue that if you are in a place where you are thinking in terms of diminishing returns, you have already hit a plateau in your organization's progress. The next set of thoughts for us is to go beyond that "do less bad" thinking. We have to find ways to use our business to "do more good." This is about getting to solutions that create more positive benefits.

Pagitsas: We often celebrate successes but sometimes do not share lessons learned when things go wrong. Have you had an experience where a sustainability goal or activity went wrong or didn't hit the mark?

Hagen: I have so many. [laughs] We missed our first carbon goal the first year that we put it together. It was by a couple of percentage points, so not by a lot, but we recognized that we didn't have our hands around all the moving parts that were contributing to our climate footprint, from fleet operations and diesel consumption to our emerging data center business and our electricity consumption. And we had a lot of scope 3 emissions in our initial carbon number, which are tough to capture accurately. Because of the number of moving parts, we were unprepared. We didn't have the metrics or the road maps and lacked an understanding of the levers that were contributing to the carbon footprint total.

Now, the great news of missing our first goal was that we had a goal set, and it gave us the reason to get the information we needed. I give our CEO and the leadership team great credit for absorbing the hit for missing our first goal and taking the attitude that we needed to learn from our process, not punish the guilty. We haven't missed any published goals since.

Pagitsas: That is a great message for all CEOs and leadership teams when it comes to setting sustainability strategy and goals, especially for the first time. A "test and learn" approach is critical to the success of a sustainability strategy over the long term.

What happened next?

Hagen: We moved to the next stage: getting to metrics backed by a rigorous methodology and a process to govern and produce them repeatedly. We began to treat the sustainability metrics with the rigor that we treat business metrics. I think that was a huge turning point in how we do things.

Also, an important part of the next stage was when we said, "All right, we're going to stop doing this with just our heart. We're going to do it with our head and our heart." That shift enabled us to engage with our employees to talk about how they can use their day job to make a difference in the things they care about. For example, having treasury people in an energy conversation allowed us access to more sophisticated finance and analytical tools, a new way of thinking of things, and new outcomes for the business. That has been a big key to our success.

Pagitsas: Looking externally, how do you engage with customers on sustainability?

Hagen: The number-one place we've seen the opportunity to "do more good" and grow the business is by engaging with our customers. The fact that we have solved ESG challenges in business-positive ways for our business gives us the opportunity to see ways we may be able to help our customers solve similar ESG challenges in their business. It gives us permission to have that conversation because we're not just trying to sell them something, we're a peer who's solved the problem, and we're trying to join with them to solve an even bigger problem together.

One example of this is in our data center business. Because of their energy consumption, data centers can be a big part of a company's carbon footprint. Through the Clean Energy Buyers Association and its Future of Internet Power working group, we joined with our competitors, customers, key ESG data assurance folks, and leading NGOs [non-governmental organizations] like the World Resources Institute [WRI] and World Wildlife Fund [WWF] to develop an open-source protocol. This protocol allows data center operators to "pass through" the environmental attributes of our renewable energy contracts to our customers so they can claim legitimate scope 2 and 3 carbon reductions.

We are the first colocation operator to turn this protocol into a standard offer. Not only do our customers get our data center services, but they reduce their carbon footprint at the same time in a way they can report to CDP or RE100 or other bodies. We've actually gone through the whole arc, in my opinion, from energy blind spot, through this process of improving our operations, to now thinking of ways that we can join with our customers to do more good. What that has done is enabled enterprise customers of all sizes to not only benefit when they do business with us, but because we helped build it on an open protocol with World Resources Institute and others, they can ask for that same thing from every data center supplier and help change the industry.

Now, the whole industry can help make the grid greener for everyone because there is power in the marketplace trying to solve the problem. We've harnessed this great lever of enterprise to solve a big challenge for everyone, which is having a carbon-free grid because we thought through how to take what we've learned and apply it to what our customers need.

Pagitsas: How do you anticipate businesses will lead and operate in five years as a result of integrating ESG principles and sustainability principles into their core products and services?

Hagen: I believe the notion of "sustainable business thinking" is the equivalent of a disruptive technology to business. It's changing everybody's job and every business, and if it's not changing your business, you're not paying attention.

The expectations for what business does and how we do it have moved dramatically.

ESG or sustainability is affecting every discipline in the business from marketing and finance to procurement, real estate, and product development. In marketing, you need to know what greenwashing is and how to authentically represent conversations. In finance, you need to know what *green bonds* are and what they mean. In procurement, you have to know what a life cycle assessment looks like and what's the resilience of your supply chain. In real estate, you must know how to develop and manage green buildings because they're better, have lower costs, and are more resilient. In product development, the ESG attributes of the product are going to become more and more important to solving customer pain points and developing better, faster, cheaper products.

Everything is being disrupted, and it's not the Patagonia model. It's the better, faster, cheaper model. Patagonia is arguably a sustainability-oriented business or brand, but for all their success as a brand, their products arguably are not functionally better. I'm not paying a premium for their coat because it's warmer or more waterproof. I am paying a premium because I believe in their brand, and I believe in their company. I want them to do good things.

That's not going to be enough in the next chapter of sustainable business progress in the conventional business-to-business space. Yes, we have to run a better operation with better environmental, social, and business outcomes, but that's not going to be enough. We're going to have to make better, faster, and cheaper products because they're ESG-advantaged. Sure, we solve customers' co-located data center needs at competitive pricing with great service. But because of our commitment to reduce our carbon footprint through renewable energy solutions, we realized that if we did it right, we can certify that and pass through the benefits to the customer, reducing their data center's carbon footprint instead of increasing it. That's just a better, faster, and cheaper product. I think that is the disruption that's going on and what's changing over the next few years. Different industries will feel it in different ways at different times, but this is what I see ahead for businesses.

Pagitsas: How do employees factor into this disruption and change?

Hagen: There are quite a few studies that say younger people want to work where they feel they make a difference. I do not think that is exclusive to young people. No one who has a choice is willingly going to leave their values in the parking lot when they come to work. That's just not okay. It's part of how this new way to think about business is blowing up "either/or thinking." There was a time when if you wanted to make a difference in the world, you certainly didn't think joining a corporation was the answer. But what if we can use our day jobs to make a living and a difference at the same time?

Our employees are probably the number-one vector for change. What we're starting to see is that offering people that opportunity drives change, delivers results and engages people by encouraging them to bring their whole selves to the challenge. They succeed, the business is better, and we're accelerating the positive impact we can make.

Pagitsas: How can employees expand their sustainability knowledge?

Hagen: I'm seeing folks from all disciplines start to add sustainable business skills and competencies to their personal tool kit. They are finding more meaning in their job, and they are finding more success. In some cases, this is formal education, such as through an MBA or a graduate certificate in sustainable business, which many universities are offering. In other cases, it's specific credentials such as for the LEED [Leadership in Energy and Environmental Design] or BREEAM [Building Research Establishment's Environmental Assessment Method] certifications. And sometimes, it's learning as they go through professional organizations like Net Impact. The common thread I'm seeing is that there is now a mature body of literature in the sustainable business space. Folks are not going to be able to accelerate with on-the-job training as fast as folks who seek out some formal process to accelerate their skills and their career.

Pagitsas: Some people get frustrated by sustainability initiatives, saying there is no way to change the negative impact that businesses are having on climate and society. How would you respond?

Hagen: If we see it as denial or despair, then we're never going to get to our hope. We could focus on the downsides for business, and the risks are glaring and scary. But there is a bigger opportunity for business which is to be better at business, to deliver better products and services, and to be more profitable, while delivering on more environmental and social benefits as we use the enterprise to drive change. That's hope.

To paraphrase a thought Paul Hawken offers in his seminal book *The Ecology of Commerce* [(HarperCollins, 1993; Revised Edition, Harper Business, 2010)], business is the only human institution with the speed, the scale, and the resources to make a big enough change fast enough to matter. That's my soapbox for the day.

Energy, Building Materials, and Equipment Manufacturing

Katherine Neebe

Chief Sustainability Officer and Vice President, National Engagement and Strategy, Duke Energy; President, The Duke Energy Foundation

Katherine Neebe is vice president, national engagement and strategy and chief sustainability officer at Duke Energy, and president of the Duke Energy Foundation. Duke Energy is one of the largest energy holding companies in America and provides electricity to 7.8 million customers in six states, serves more than 1.6 million customers with natural gas in five states and through its commercial business owns and operates diverse power generation assets, including a portfolio of renewable energy projects. Katherine leads Duke Energy's strategy to deliver on the dual goals of meeting customer needs for reliable and affordable energy and achieving the company's goal of net-zero carbon emissions by 2050.

Katherine's deep expertise in stakeholder engagement, environmental sustainability, and social impact is grounded in her prior roles at Walmart and the World Wildlife Fund (WWF). Her work has spanned leading Walmart's environmental, social, and governance strategy (ESG) and managing one of the world's largest corporate non-governmental organization (NGO) partnerships with The Coca-Cola Company, focusing on water, agriculture, and climate in more than 45 countries.

© Chrissa Pagitsas 2022
C. Pagitsas, *Chief Sustainability Officers At Work*,
https://doi.org/10.1007/978-1-4842-7866-6_14

Katherine was a First Movers fellow through the Aspen Institute in 2019.

Chrissa Pagitsas: Duke Energy plays a critical role in the US utilities infrastructure. Can you tell me what services and operations Duke Energy encompasses?

Katherine Neebe: Duke Energy is one of the largest energy providers in the United States of America. We operate a vertically integrated utility, which means we have generation, transmission, and distribution. There are around 30,000 employees who work for the company. We serve around seven million customers, primarily in the Southeast and Midwest regions. We have been in business for over 100 years.

Our corporate purpose is to power the lives of our customers and the vitality of our communities. Our business goal is a commitment to achieve net-zero carbon emissions by 2050. This commitment has two interim milestones—to be net-zero in methane emissions by 2030 and to at least halve our emissions from electricity generation by 2030 relative to a 2005 baseline. Fundamentally, I would argue that one of our biggest challenges as a society is climate change, which drew me to Duke Energy. The company has other products and services, but how we solve climate change is imperative.

Pagitsas: Your role is vice president of national engagement strategy and chief sustainability officer. You are also president of the Duke Energy Foundation. What do those roles encompass?

Neebe: I was brought on to focus on three primary areas at the company—sustainability, stakeholder engagement, and our philanthropic work. Specifically, I lead Duke Energy's stakeholder engagement efforts to develop solutions to meet customer needs for continued reliable and affordable energy while simultaneously working to achieve the company's goal of net-zero carbon emissions by 2050. In my capacity as president of the Duke Energy Foundation, we provide more than $30 million in support to meet the needs of communities where Duke Energy customers live and work. 2021 has been particularly challenging, and I'm proud of the way the Foundation rapidly pivoted to address both COVID-19 relief and racial equity and social justice efforts.

Pagitsas: Why is sustainability important for Duke Energy?

Neebe: We are at an interesting time for the energy sector because it's going through a massive transformation. When I think about Duke Energy's ability to achieve its ambition by 2050, the pace of change will be tremendous. We have a clear line of sight to the technology required through 2035 as we seek to decarbonize the grid.

Many of the technologies that will take us further, all the way to net-zero, are in their nascency but rapidly advancing and hopefully coming to scale soon. When the sector transforms and decarbonizes, we will be able to make a substantive contribution to help mitigate the impacts of climate change in a way that a lot of other companies simply can't because they're not in the sector that we're in.

Pagitsas: Where is Duke Energy on its sustainability journey?

Neebe: We're contemplating the largest retirement of coal-fired units in our industry, and we have already retired fifty-four coal-fired units. This is a journey that we've been on for over a decade, and we've already seen a lot of success. We've reduced our emissions by more than forty percent relative to 2005. So, we know what good looks like. But we have work in front of us as well. It's exciting and will be challenging. It will require great dialog, partnership, and conversation with all of our stakeholders, including our regulators, our investors, our customers, and civil society, to get to net-zero in a way that maintains reliability and affordability.

Pagitsas: What additional impacts does Duke Energy's strategy have?

Neebe: Sustainability is not just the environmental side of the table. It also looks at economic and social impacts. When I think about social, I think about three areas of focus.

First, I always start with what's in our sphere of influence. So, how do we think about jobs and the transformation? We know that technologies are going to evolve, and the job requirements are going to evolve. We're thinking about how we are preparing our workforce by up-skilling and re-skilling and doing things of that nature. That's one way we are looking at the intersectionality of social and economic impacts.

I also think about our customers, a second core area of focus. For example, let's consider the COVID-19 pandemic. We know our customers and our communities were impacted profoundly by the pandemic in terms of job loss—in particular, those who are underserved. Underrepresented communities have been disproportionately impacted. As a utility providing power, we have to think about all of our customers and communities and how we are serving them and evolving our business to meet their needs.

Then the third core area is this notion of community vitality. I think about the role that Duke Energy plays in our communities, not just from jobs but from sourcing and how we're bringing other companies and other industries to the regions where we operate. Together, we are helping to bring new jobs, economic growth, and economic development. I think that's critical.

Recently, we had a team meeting where we invited the head of supply chain to share a little bit about what the team is doing and some of our work to focus on hiring in North Carolina—so, sourcing directly from our headquarters in the state of North Carolina—as well as spending north of a billion dollars with tier-one and tier-two diverse suppliers, including women-owned and minority-owned businesses. In other words, our business strategy of net-zero manifests in issues beyond climate. And I try to think through all of the dimensions as it relates to environmental, social, and governance principles and how we are applying them to our work.

Pagitsas: Duke Energy's environmental strategy seems to be intricately woven with serving its community. Why is that?

Neebe: Yes, it is. The unique thing about Duke Energy relative to every other company I have worked at or with is that you cannot tease out our sustainability strategy or our climate strategy from our business strategy. At some companies, you sometimes try to figure out how to make sustainability relevant. In some cases, you're trying to shoehorn it in because it's new, and you're trying to figure out how to make it work and embed it across the company. In other cases, you're trying to shoehorn it in because it doesn't really fit, and you're trying to make a case. I think what's different at Duke Energy is everyone at the company is playing a role in our climate strategy. It's our business strategy. We're making a $59 billion investment in our climate strategy over the next five years, and that's our investment in the future growth of the business.

Pagitsas: Do you have key partners sitting at the senior leadership table who are invested in the success of the sustainability strategy?

Neebe: Yes, and it's not just the key leaders. I would argue every person at Duke Energy has a role to play in delivering that promise and our net-zero strategy, which is so cool.

Pagitsas: What are some examples of Duke Energy's business lines engaging that align with Duke Energy's sustainability strategy and goals?

Neebe: There's a fairly straightforward example of those involved in siting, building, maintaining, and scaling renewables within our regulated and commercial business. In fact, our commercial renewables business includes utility-scale and solar generation assets, distributed solar generation assets, distributed fuel cell assets, and battery storage projects across twenty-two states from twenty-one wind facilities, 150 solar projects, and two battery storage facilities, and we plan to further grow our commercial renewables portfolio by 2050. This is one critical piece of the puzzle. But I would also argue that meeting our net-zero goal includes the teams on the transmission and distribution side of the business who are playing a critical role in strengthening our grid to enable that renewable energy to go where it is needed.

In another case, we have teams that leverage technology to do things such as help customers use energy more wisely or improve restoration times in the event of severe weather. In that latter case, smart meters play a key role in improved restoration times. Last year, restoration teams were able to successfully ping 41,000 meters to verify that power had been restored after completion of repairs, saving around 5,500 truck rolls and freeing up resources during major event responses in 2020.

Pagitsas: On the environmental side, what is an example of Duke Energy's investment to meet that ambitious 2050 goal?

Neebe: Yes, let's talk about how we're going from point A to point B. Just for clarity, we've already reduced our own carbon emissions by more than forty percent since 2005. Many people may not know or fully appreciate that we're already a leader in low-carbon intensity. I point to the Carolinas, where fifty percent of the energy produced is from a zero-emitting source, which is nuclear. As it relates to energy intensity, we're already a strong company and a leader in our industry.

Of course, we're overseeing the largest coal retirement in our industry with a plan to retire all coal-only units by 2030 in the Carolinas. We've accelerated the closure of coal plants in Indiana, shortening the average retirement dates by about forty percent on top of the fifty-one units we've already retired across our fleet. So, this is a transformation that is not new to Duke Energy. It's been a transformation journey we've been on for quite some time.

Pagitsas: What's the strategy behind the transformation journey?

Neebe: There are four things that I'll dig into and unpack. The first thing we know is that we need to invest in more renewables. We've got to focus on expanding energy storage. We also know there is a role for carbon-free nuclear. I spoke about it before in the powerful role nuclear energy is already playing today. Then, we need to think about dispatchable resources which is energy that power grid operators can send on-demand according to market needs like natural gas. Those last two technologies I spoke about when it comes to affordability and reliability cannot be understated.

So, on the renewables front, today, we've got around eight gigawatts that we've contracted, we own, or we operate. We are on track to pass 16,000 megawatts by 2050 and 24,000 megawatts by 2030 for our regulated utilities. By 2050, renewables will represent forty percent or more of our energy mix, which is significant growth.

On the storage front, a lot of attention is being placed there. During the next five years, we anticipate spending around $600 million on new battery storage. What I find fascinating about storage is that everyone I run into thinks that batteries are a new-fangled technology that is being developed and scaled. We are seeing real progress in battery technology and how it's helping maintain reliability, particularly as you bring on technologies like wind and solar that are more variable.

However, there's battery storage that's been on the grid for years. For example, Bad Creek and Lake Jocassee, hydroelectric stations owned by Duke

Energy, are pumped hydro storage and represent more than 2,200 megawatts of power. This technology is more than 100 years old. While there are a variety of different battery technologies, we expect to grow battery storage by over 13,000 megawatts by 2050. As a technology, battery storage is a great example of innovative new technology and technology that's been around for a long time.

Pagitsas: You've mentioned innovation. Some would say that utility companies are not innovative and don't use new technology. Yet you're saying something different. You're saying that there is "old" technology that's good, and then there's new technology to be explored. Why are you using "old" technology?

Neebe: In my experience, the sustainability community tends to chase after the bright and shiny or the next new innovation that's going to save us from tomorrow but isn't yet available. I am sympathetic to the argument that innovation is going to play a critical role because we do need it. But I think we need to look at the technologies that we have on hand today and figure out how we fully utilize them. Are they fully leveraged? Are we applying them in the most meaningful way? And I get the appeal of the sexy new technology, but we have to realize what we can use here and today. So pumped hydro storage? It's a technology that's working today. Energy efficiency and daylighting for buildings, for example, are also available today. These are all an important part of the mix and have a role to play in our energy future.

I also look at nuclear energy. Nuclear provides around fifty percent of the power we generate in the Carolinas. There is new nuclear technology on the horizon, which is one of the potential technologies that we refer to as ZELFRs, or zero-emitting load-following resources. These are important because as you bring intermittent energy and distributed energy on the grid, you have less control of those sources as a utility. In other words, you are dependent upon outside and external factors such as the sun shining and the wind blowing. To maintain reliability, you need a dispatchable fuel source, and you can bring it on—or down—quickly. Examples of ZELFRs include advanced nuclear, hydrogen, long-duration storage, and carbon capture, utilization, and storage technologies.

If we balance what we have on the grid today that is working with some of these new technologies that will be scaled and market-ready at some point in the future, I think that mix is really exciting. That's how we're going to deliver our net-zero ambition by 2050. It's by using technology today and the technology of tomorrow together.

Pagitsas: How did you start your career in sustainability? What have been the key milestones along the way for you?

Neebe: It's funny how my career always makes sense in hindsight. In the moment, it sometimes seemed less strategic! I got my start in 2000 when I worked for a small consulting company in Oregon that received some of their funding from utilities' public funds to encourage people to install energy-efficient appliances into their homes. It's a full circle to my work at Duke Energy!

At the consulting company, I worked on getting people to buy ENERGY STAR products and appliances and install them in their homes. I worked with a lot of utilities that were very interested in demand-side management. I also worked with those in civil society who were interested in reducing emissions and with other organizations that were focused on helping people with their utility bills. Finally, I worked with manufacturers and retailers who were interested in new market opportunities and selling a product that had a higher margin.

I just fell in love with the sustainability field, which, in my view, is about how we incorporate environmental and social objectives into the business world. It's also the opposite of that. How do you incorporate business into environmental and social challenges? Fast-forward a bit. It took me a while to find my way in the field of sustainability. I went to the University of Virginia Darden School of Business and got my MBA because I thought I would stay in this field that was a little bit on the fringe. I needed to understand how to make sustainability effective within a business.

From there, I spent about a year focused on hog farms in eastern North Carolina and working on sustainable agriculture, an issue with ties to environmental justice. Then I moved up to Washington, DC, where I worked as a contractor for the EPA [US Environmental Protection Agency], looking at things like clean indoor air. At the time, it was kind of a patchwork of jobs that I wasn't sure would come together to make a career that made a lot of sense, but I was intrigued by it. We're talking about all this happening in 2003, 2004, and 2005. Sustainability was still new, very much nascent to the private sector and before they embraced it.

About that time in 2004, I was back home working on sustainable agriculture in eastern North Carolina, and one of the people I bumped into was Hilary Davidson. She talked about climate change and worked at Duke Energy. I remember at the time being so impressed because there were so few of us working on sustainability, but to see someone from a utility that provided the power to my parents' home talking about climate change was really awesome.

In 2007, I received an outstanding opportunity to work for the World Wildlife Fund, running a very large partnership with The Coca-Cola Company looking at freshwater, which is the main ingredient in all of Coca-Cola's products. If they don't have a sustainable water supply for their manufacturing, and if the local communities around those plants don't have a sustainable supply of water, they have some real business risks. That focus has helped them take on initiatives like water efficiency, climate, and sustainable agriculture and create a best-in-class sustainability program on an issue that is core to their business success in the long term. It was extraordinary to work on that.

From there, I moved over to Walmart for seven and a half years. My career evolved in the years that I was there. From stakeholder engagement to working on particularly complex sustainability challenges like human trafficking in the seafood sector and human rights to figuring out how the company plays a role in what I call "international policy," specifically the business voice at the UN COP [Conference of the Parties] for climate—so Paris and Bonn, and then some C-suite support on sustainability projects.

Most recently, at Walmart, I led ESG for the enterprise. I tackled the environmental, social, and governance issues that the company needed to understand were coming, how to close any performance gaps relative to those issues, and how we told our story to investors and other stakeholders, particularly thinking through the emerging mainstream investor conversation around ESG. I did that for about two and a half years. Now I'm at Duke Energy, where I have sustainability, stakeholder engagement strategy, and the Foundation. As you can hopefully see, my career comes together, but in hindsight.

Pagitsas: Clearly, you have been in this sustainability field for a long time and have seen it evolve. How would you describe it for somebody trying to wrap their arms around sustainability versus ESG? What about corporate social responsibility and impact investing? Those are some of the keywords we hear in the news these days.

Neebe: I don't spend much time getting into the debate about which one means what. This is a field where the language evolves, the trends evolve. What I always come back to is the standpoint of a business. Fundamentally we're talking about how we're mitigating risk for environmental and social issues and how we're creating value on dimensions of environmental and social factors. I set aside governance only because I think that's fundamental to any business—governance, good ethics, compliance, and a great culture infused with values and purpose. I think good governance is fundamental to any strong business. I would set that as a distinct issue relative to ESG or sustainability.

When I advise people who are new to the field, I always suggest starting with two questions for the business. What issues matter to the company? To which issues does the company matter? These are two questions about materiality, specifically as considered by sustainability practitioners. How do you manage that risk? How do you create value? You can call it corporate social responsibility. You can call it sustainability. You can call it ESG. You can call it whatever you want. But fundamentally, what you are trying to do as a company is figuring out where you need to move and what you need to address.

Pagitsas: Let's discuss leadership. You've sat in many board meetings, and you're a part of the executive management. What are the key skills that you apply in your jobs that you think have made you successful as a sustainability leader within business?

Neebe: The things that are most fundamental are listening, prioritization, and then creating positive impact, with a subtle point about getting over the *no*. Taking a step back, sustainability broadly defined is problem-solving and change management. In my experience, when you want to work on something, and you're trying to convince someone, the best thing you can do is listen to them. Understand their values, and understand their challenges. More often than not, someone will tell you not only why they can't do something but what they need to overcome. There's too much telling in this field and not enough listening. So that would be one.

Prioritization, second, is so critical. I go back to the questions, "What issues matter to the company?" and "To which issues does the company matter?" as being instructive in how you prioritize. There are a lot of people coming to companies with a lot of ideas, a lot of solutions, and a lot of priorities. For companies to be really effective, they must focus on that prioritization. It's essential for them to knock out what they fundamentally can and should before they take on more. Again, this is a field where so much is coming at you.

The third skill is creating a positive impact. The way I've chosen to define impact is to work with very large corporations and help them pivot and evolve their business strategy in a way that captures value for taking action on sustainability. I would argue what draws people to this field is this question of impact. There are a lot of different ways that you can define impact. As a leader, throughout your career, it's important to define or evolve what your own personal definition of *impact* is.

Lastly, there is getting over the *no*. This is a marathon, not a sprint. So, understanding when a *no* is a *not now*, and when it is a *not ever* or a *no, I'm scared*. I would go back to that first point to listen, but the subtle art here is to try to figure out how you get someone to a *yes*.

Pagitsas: It seems what you're describing, the getting over the *no* would be an under-the-radar skill. You're not going to see it on the job description, but it seems to be critical to the recipe for success as a sustainability leader.

Neebe: I talk an awful lot about changing hearts and minds and how you can get people to think differently. It's not about me, the practitioner, having a good idea. It's about how you're transforming every organization you touch— ask smart questions and understand the nature of risk and value.

When I was at WWF, we did a site visit at a Coca-Cola manufacturing facility or production center. This was back in the mid-2000s. I was the NGO representative, and this was a time when it was novel, scary even, to let an NGO, even from WWF, into a facility with the plant managers. And we would listen. We would ask, "What you are making? What are your business goals and challenges in this region, in this country?" We'd ask for a tour of their facility. I wanted to see how they manufactured a Coke beverage from beginning to end.

We would try to conclude the visit by asking, "Where does your water come from?" More often than not, they'd point us to the water pipe coming into the facility. And we said, "Great. For the next portion of our time together, let's go to the watershed where the water comes from." This would allow them to look at the dimensions of risk differently from the standpoint of the risk that the watershed is facing. That part of the facilities visit helped change the way someone thought about risk and opportunity. It was fundamentally transformative. I think the secret sauce of effective sustainability is that transformation.

Pagitsas: Any other under-the-radar skills for sustainability professionals preparing to enter the field or propel their company forward?

Neebe: I think the exciting thing about sustainability is that you almost have to be a futurist. I like that the field is about looking around corners at what's coming. What are the trends? What are the issues? What are the risks? What are the opportunities? And are they relevant to the enterprise? If so, how? Where's that opportunity, and where is there that risk? Are you teeing up the business to be responsive when that thing in the distance is suddenly in the here and now? There's a blend of forward-looking thinking needed with a critical eye on how, where, when, and why it will manifest itself in the company. That is what's important—and then being able to articulate why that future issue matters.

The other thing that is often underappreciated about sustainability programs or ESG programs is we are often the advisors to the business. Now there's some work that we own, and certainly, there's work that we embed into an organization. However, it's the internal business decision-makers, not the sustainability team itself, who are fundamentally transformative.

This is a conversation that I often have with people who want to get into the field of sustainability. There's a perception that unless the word *sustainability* is in their title, they're not empowered, and they won't be able to make sustainable business decisions. Whether I was working at Walmart, advising Coca-Cola, or now stepping into my role at Duke Energy, the business decision-maker decides what to source, where to put technology, or what to invest in. That is the person that is truly making the transformative change.

Pagitsas: One of the topics that you touch on is the transformation of stakeholders toward sustainability. What does stakeholder management mean to you?

Neebe: I think business has a unique opportunity right now given where society is to fundamentally ask itself, "What is the role our services or products are providing society?" The question I'm intrigued by is, "How is business serving society?" That, to me, is the hardest stakeholder engagement question.

One of the other reasons I was attracted to Duke Energy, Walmart, and working with Coca-Cola was fundamentally to ask the question, "What societal role is this company playing?" Duke Energy is providing power. We have an obligation to provide power to everyone in our service territory. That in and of itself is a societal good. Being able to turn on the lights, power your refrigerator, or run a business is enabled by our company. This is fundamentally a societal good. The ability to do all of that reliably and affordably is also critical. Now we're layering in increasingly cleaner energy as we address climate change.

When I think about the way that we're solving societal problems or our stakeholders' problems, I'm inspired every single day. It inspires me when I think about what this company can do to advance issues like diversity, equity, and inclusion, how we address climate change, mitigation, adaptation, and resiliency, and how we help build strong economies as a business through philanthropy and with our stakeholders.

When I say "stakeholders," it could be civil society, the customer, our employees, the community, or investor—basically, all the dimensions we serve. Fundamentally, it's about getting people together to ask questions such as, "What outcomes do we aspire to?" and "How do we optimize the system so that we can get to those outcomes?" And another important question is, "What is the cadence of change that is necessary, and what must that transformation look like?"

At Duke Energy, the specific questions I am asking are aligned with, "How do we work together to figure out what are the outcomes we aspire to?" and then, "What is the role for Duke Energy in those outcomes?" That, to me, is when you're doing best-in-class stakeholder engagement. That's what it looks like. That's the conversation you're having. It's not stakeholder *engagement* or *management*. It's about *outcomes*.

Olivier Blum

Chief Strategy & Sustainability Officer
Schneider Electric

Olivier Blum is the chief strategy and sustainability officer at Schneider Electric, a multinational company with revenues of more than 25 billion euros in 2020. With headquarters in the United States, Asia, and Europe, Schneider Electric provides solutions in the digital transformation of energy management and automation bringing efficiency and sustainability within homes, buildings, data centers, infrastructure, and industries. Corporate Knights recognized it as the most sustainable corporation in the world in 2021.

Olivier leads the development of Schneider Electric's strategic and sustainability initiatives and merger, acquisitions, and divestment activities. Before taking on the role of chief strategy and sustainability officer, he was the global chief human resources officer. He also held senior positions in China, India, and Hong Kong, including regional head of strategy and marketing director, regional managing director, and executive vice president of retail. He began his career at Schneider Electric in France.

In 2019, Olivier received France's Chief Human Resources Officer of the Year Award from Cadremploi, Morgan Phillips Hudson, Le Figaro Décideurs, and Fyte, in recognition of how he transformed Schneider's leadership and culture at a global scale. He currently serves on the Delta Dore board and on the AVEVA Group PLC board as a non-executive director and remuneration committee member.

© Chrissa Pagitsas 2022
C. Pagitsas, *Chief Sustainability Officers At Work*,
https://doi.org/10.1007/978-1-4842-7866-6_15

Chrissa Pagitsas: You have had a unique career path, Olivier. You started your career at Schneider Electric and have been there since 1993—or twenty-eight years as of 2021. The first C-suite role you held was chief human resources officer [CHRO]. Now you concurrently hold the roles of chief strategy officer [CSO] and chief sustainability officer [CSO]. Effectively, you're a "double CSO." What career path within Schneider Electric led to your leadership in these different executive roles?

Olivier Blum: An interesting characteristic of my career is that I've been in a lot of different roles—from the commercial business side to product development both at the global and local level. I have had the chance to discover all the dimensions of the company. The second interesting characteristic of my career is that I have worked internationally for many years. Ten years in Europe and eighteen in Asia. Right now, I'm located in Hong Kong. I spent almost six years in China and six years in India, which has probably changed a lot of my vision of the world, leadership, and diversity and inclusion.

I grew up in France and joined Schneider Electric after graduating from university, so I have spent my entire career at Schneider Electric. I first joined the sales and marketing group, and then for the next seven years of my career, I worked in the commercial division. In 2001, I was selected to become the chief of staff to our then CEO, Henri Lachmann. It was unique for me to assume that role because I was in my thirties—quite young, and I did not know much about how a global company works.

As chief of staff until 2003, I was responsible for leading the company's transformation program. That's the first time we decided to put social issues on the company's agenda. In 2003, I moved to China, and I have now spent most of my career in Asia. In China, I led the Final Low Voltage business unit and then later strategy and marketing. In 2008, I went to India to oversee the market after which I went on to lead the home and distribution division globally.

In 2014, Jean-Pascal Tricoire, the CEO of Schneider Electric since 2006, wanted to accelerate the leadership and cultural transformation of the company. He offered me the CHRO role for the group, which focuses on people, diversity and inclusion, and social issues, which are also very important to a sustainability strategy. Most recently, in 2020, I moved over to lead the strategy and sustainability roles. I am now based in Hong Kong.

Pagitsas: Until 2014, your roles at Schneider Electric were leading businesses in China and India. How did your CHRO role come about? What was your view of the business and cultural transformation underway?

Blum: When you are a business leader, you must lead your business and achieve your strategic objectives. But at the end of the day, fifty percent of your job is about the leadership of people. It's about the people you work with, the people you hire, people you develop, and the type of culture you develop in the company.

When Jean-Pascal and I discussed the CHRO position, he said, "Look, I think we are about to succeed in our business transformation. Historically, we were a medium sized product company moving to become a large solutions company that will be part of the solution to the clean energy transition. We have diversified our business. We are moving in the right direction, but the company's culture and leadership are not moving as fast as the strategic transformation."

While I'm not an HR professional by training, I was quite intrigued and excited by the challenge. Every day, HR should have a strong impact on culture and leadership and not only do the transactional work. I was excited to lead this transformation at scale for a company the size of Schneider. When you have been challenging a function for many years, and you are given the position to be part of the solution, it's difficult to refuse.

Pagitsas: How did your international experience prepare you to be a CHRO and think about diversity?

Blum: Before I went to China in 2003, I could check off successes on my résumé. I'd gone to a good European university, been successful in the company, and worked with the CEO. Then, I went to a country which was so different from what was known and familiar to me. There was a lot I didn't understand. The Chinese market was very different from everything I'd learned so far in my life in Europe. That was a very challenging situation.

That's when it clicked, and I realized that the world is not exactly the way I had looked at it so far from my European angle, education, and background. It was then that the journey about the importance of diversity within a company probably hugely accelerated for me. I began to understand that talent is a rare resource that you need to attract. It was very difficult to attract the talent necessary for our business because it was not available in China at the time. You needed to develop and retain the talent.

When I moved to India six years later, I thought it would be easier but one month into my stay, I realized there is a lot I don't understand about the country again." My way of looking at business and diversity changed a lot because of those years in China and in India.

Pagitsas: Let's next look at your chief sustainability and chief strategy officer roles. Why were these roles combined? What are the synergies?

Blum: Historically, the topic of CSR [corporate social responsibility] was treated separately from the rest of the business' activity, but that has changed over the years. The willingness to make the world greener and more equitable is not just a moral responsibility. It makes good business sense too. Sustainability is at the core of our purpose and mission as a company. I lead both strategy and sustainability because sustainability is deeply ingrained is in our strategy.

The scope of the chief sustainability officer is divided into two parts. First, it includes everything we do on our own internal sustainable or environmental, social, and governance [ESG] journey. Second, we help our clients focus on the environmental part of their ESG strategy. Our mission is to be the digital partner on energy efficiency and sustainability for our customers.

Pagitsas: It seems there have been two transformations at Schneider Electric. One has been a shift of focus from products to services, and a second has been the embedding of sustainability into the core strategy.

Blum: That's true. When I was appointed as CHRO, we as a leadership team asked ourselves, "How can we make this company different? How can we make it unique with a positive impact?" Our whole approach is to be a progressive company, permanently push the frontiers, and innovate to have a positive impact.

At Schneider Electric, we like to say that sustainability is like a marathon, even a race without a finishing line. Because there are no limits to the positive impact, you can have when it is in everything you do. That's why people join Schneider Electric and stay, especially the leadership. It's because we're always trying to push to the next level. We are always asking ourselves, "What else can we do to have a better impact on the people, the planet, and our customers?"

Pagitsas: Is this long-term race approach different from earlier iterations of sustainability and CSR at Schneider Electric?

Blum: Twenty years ago, around the early 2000s, it was more about philanthropy. Then fifteen years ago, around 2005, we started to pick up the concept of sustainability. It became very core to the culture of Schneider. For us, sustainability is not about being compliant with ESG reporting and disclosure. Yes, it's great to do it, but it's more about what else you can do to have a positive impact as a multinational everywhere you operate in the world.

Pagitsas: What is the sustainability strategy that you've put in place now, Olivier, and why is it important for Schneider Electric to achieve those goals?

Blum: Schneider Electric's purpose statement is to "empower all to make the most of our energy and resources, bridging progress and sustainability for all." Our purpose is to help our customers be more energy-efficient and go through their own decarbonization journey.

Schneider Electric has set its strategy around six pillars—Climate, Resources, Trust, Equal, Generations, and Local. The Climate pillar's scope is about progressing fast to net zero in our operations with partners and for our customers. The Resources pillar addresses advancing circularity and preserving our planet's riches and biodiversity. Trust is important to our culture because it upholds employees and partners to our social, governance, and ethical standards. The Equal pillar focuses on supporting diversity and inclusion at work and those with no access to energy. We are committed to upskilling and developing everyone across generations to contribute and thrive as part of our Generations pillar. Last but not least, our Local pillar is about empowering local teams and partners to lead and deliver grassroots impact. The topics of climate and resources are supercritical because we want to be role models to our customers in everything we do as a company.

Pagitsas: The pillars cover a lot of ground. Let's drill into a specific goal within one of them. Under the climate pillar, you have a goal focused on your suppliers. The goal is to reduce carbon dioxide emissions from the top 1,000 suppliers' operations by fifty percent. That's an ambitious and large goal. How are you going to achieve this goal, Olivier?

Blum: We started to look at our carbon dioxide emissions many years ago. We started to measure carbon emission at Schneider Electric in 2006, and now we are quite advanced. Today, if you look at our carbon footprint, our scope 1 and scope 2 emissions, which are our operational emissions, are only 300 kilotons of carbon dioxide, less than one percent of our carbon footprint. We have reduced our operational emissions by sixty percent in just the last three years. And very soon, we will be carbon neutral in our operations.

Our operational emissions are fairly small compared to our scope 3 emissions coming from our upstream and downstream value chain, such as our suppliers and customers. Our new strategy reduces our scope 3 emissions upstream and downstream. Over the last fifteen years, we engaged our suppliers on people issues such as human rights issues, safety in their factories, and so forth.

We have engaged our suppliers now on the carbon topic because, for those suppliers, the topic of climate and carbon emission is getting higher and higher priority for them as it becomes higher for their customers. Yet, you have many companies who don't know what to do about it. They don't know where to start and don't necessarily have the means to engage a large consulting company, and so on.

With this culture of involving our suppliers in our sustainability journey, we have engaged our top 1,000 suppliers, the ones with the largest carbon dioxide emissions. They are not the largest suppliers from the financial or product standpoint but, in fact, the ones with the largest emissions. We created a program for those 1,000 suppliers to help them define their carbon emissions baseline and go through the consulting services that we have in different parts of the world.

Pagitsas: What is an example of how you have helped suppliers reduce their carbon emissions?

Blum: For instance, the topic of purchase power agreements [PPA] for renewable energy is very important. The electricity consumption of some individual suppliers might not be high enough for them to access a renewable PPA. Hence, we aggregate demands of various suppliers for collective contracting of renewable energy.

We launched the program in March 2021. It was very interesting to see that ninety-nine percent of the people who have been invited participated in the program launch webinar. After one day, we had twenty-five percent of our suppliers subscribe to the program, and within a few months, 100 percent of our top 1,000 suppliers had joined the program.

Pagitsas: Why were your suppliers interested in enrolling in the program?

Blum: It's interesting because investors asked us the same question when we discussed this goal and program with them. Carbon emissions and climate change are important for the suppliers. When you can follow a leader such as ourselves, if I may say, and get the benefits from our own expense and services, it's the ideal situation for them. We are in year one, so we are beginning by helping them to measure their footprint. For those who have already measured their emission in the past, we're helping them do the next step, which is develop the carbon emissions reduction strategy. Progressively, the objective is to deliver on their carbon emissions reduction program for the next five to ten years.

Pagitsas: Is helping those suppliers reduce their emissions a strategic move to reduce the risk for Schneider Electric's supply chain?

Blum: Absolutely. It has multiple advantages. First, it's about de-risking your supply chain because those companies will have to go down their own decarbonization journey anyway. They will have to provide materials to Schneider Electric that are less and less carbon-intensive, whether steel, electrical feed, or electrical vehicles. Second, while it's about de-risking their business, it's about also reducing the carbon dioxide intensity of everything they sell. Third, since we are a provider of carbon emissions reduction solutions to the supplier in some cases, it's also a business opportunity for us.

Pagitsas: Let's dive into another one of the pillars of the strategy in more detail. Which one would you like to explore?

Blum: The "Equal" pillar of our strategy is about diversity and inclusion. We've been working closely with our CEO to cultivate a diverse workforce and create an inclusive culture. Bottom line, we want to create equal opportunities for all because we are obsessed with diversity at Schneider Electric, whether it's having more women in management, more nationalities in our leadership, or ensuring equal opportunities for everyone no matter their background.

We love the idea of creating equal opportunity for all everywhere in the world. We even changed Schneider's HQ [headquarters] model in 2012. We no longer have one central HQ in Europe. We have HQs in France, the United States, China, and India. Basically, Schneider Electric's top jobs are divided into different geographies.

Pagitsas: It's a unique leadership strategy.

Blum: Yes, it is. If you look at the typical multinational company, usually they have one center of gravity. This is where you have ninety percent of the global jobs. And guess what? Because you have 90 to 100 percent concentration of the global jobs in one location, most of them come from that country. Therefore, American companies are very American. French companies are very French.

We decided to split the executive committee between the United States, Hong Kong, France, and India. Following that, we looked at the company's top 500 global jobs. We decided to relocate these top global jobs to different geographies in Europe, Asia, and the United States. It's not about sending more expatriates to Asia and United States but making sure we use local talent in each geography. Our target is to have eighty percent of the people in each HQ be from the HQ's region. We are going through this journey, we are not yet there, but we have progressed.

Pagitsas: What drove you and the Schneider Electric leadership team to implement this model?

Blum: Typically, when you develop local talent at a large multinational company, people grow but cannot get the bigger, global job if they are not located at the headquarters. Many people are not open to international mobility because of several personal reasons, such as caring for parents.

If you want to retain and develop your best talents, you have to develop them locally and give them opportunities there. We have delocalized the top jobs at Schneider to make sure that we can get a higher proportion of non-French and non-Europeans into leadership. It has had a double impact which was not necessarily anticipated at the beginning of this strategy.

Pagitsas: The first impact is having more leadership from individuals born and raised in each country. What was the second impact?

Blum: Yes, that's the first impact. The second impact is related to women. When you think about the women in leadership, many women are quite successful, but they are even less mobile than men for plenty of reasons that you can imagine. Usually, it's because they are caring for their family. It's true in Europe. It's true in North America. It's even more true in Asia, where usually the female in a couple is taking care also of the elderly parents. With our multi-HQ model, we have not only realized that it had allowed us to have more nationalities represented, but it has helped us to have much more female leadership.

An example of the success of our approach is the appointment of a new CHRO after me. She is Chinese and located in China. She probably would have been unable to become the CHRO of Schneider Electric if the job were in France.

Pagitsas: I was intrigued by the last pillar, "local." The goal associated with the pillar states that 100 percent of country and zone presidents should define three local commitments that impact their communities in line with Schneider Electric's sustainability transformation. Why was that an important goal to set?

Blum: We were one of the first companies fifteen years ago to focus on sustainability and make a global commitment. For every single strategic initiative commitment, we always say, "We are going to take the next three to four years to deploy that transformation, everywhere in the world." Remember, our goal is to make a positive impact at the global level. This global strategy has allowed us to make very good progress.

While many commitments need a global program deployment, others need a local focus. This is what has made our local pillar at Schneider, I think, very different. Let me give you an example. You look at diversity and inclusion. Consider gender diversity. It is a global topic. In contrast, Black Lives Matter is very specific to the United States. It is very important in the United States, and hence we need to have a specific focus on this topic there. Each country needs a country-specific approach regarding certain issues.

For me, what was very important is to balance the fact that taking global commitment and driving environmental, social, governance everywhere in the world is important for a large multinational. But at the same time, if you want to have a positive impact, part of your agenda has to be local.

On top of the global commitment that Schneider is making, we are asking that country leader, "What will make the difference in your country? What should you do in your own country to have a stronger and positive impact?" That's a way to rebalance a little bit and to make sure we always have this global and local dimension.

Pagitsas: Given the large global and local ambitions you've set, what keeps you up at night about your sustainability strategy? What's the thing that you want to move forward faster on but haven't been able to yet?

Blum: I would say the number one is staying ahead of the game, staying innovative, and always looking for the next frontier. We want to make sure we look at the future and what's next. The second one is making sure that we deliver on our ambitious commitments. When we build a new strategy, we do a lot of work over six months. We analyze what we are doing well, what we are doing not well, and what others are doing in the market. Then we set very big stretch goals. For example, you asked me earlier about the supplier program. If you next asked me, "Are you confident that you're going to achieve its goals?" I can't say for sure, frankly speaking. What is important is setting bold ambitions that are needed and doing your best to achieve them.

Pagitsas: Looking broadly at the field of sustainability, what are the challenges you anticipate ahead?

Blum: Today, everyone is more interested in sustainability and ESG. Therefore, what's happening? We are trying to standardize everything. For example, you have new standards being published everywhere. You have the ESG rating agencies, you have SASB [Sustainability Accounting Standards Board], you have GRI [Global Reporting Initiative], and it's all good. I'm very supportive of all this reporting because they have pushed all large companies in the world to have a strong focus on society. It's good for the planet. It's good for the people.

But the drawback of that is you're just standardizing everything. If your company stops at compliance reporting, there will be no progress or innovation as a company or society. This is because standardization reflects the past. It does not reflect the future. Let's take carbon as an example. Today, it's a topic that everyone should be focused on. But what's next? What can we do next in biodiversity? When we standardize too much of a business's activity, it becomes a kind of tick-the-box exercise, and nobody tries to be progressive and innovate. For me, again, the definition of sustainability is to have a positive impact. If you want to have a positive impact and the world is changing around you, you should change.

Pagitsas: Is there an example of standardization that supports innovation in sustainability?

Blum: Look, the very positive thing that will force every company in the world to focus on sustainability is the UN Sustainable Development Goals [SDGs]. The 2015 SDGs from the Climate Paris Agreement are good because they create a universal standards framework for everyone in the world. Six years later and in the middle of COVID-19, all companies are embracing the topic. A lot of the focus of standardization is on reporting. While we need the world to be at least at a minimum level when it comes to sustainability, innovation and sustainable innovation are the essence of sustainability. We have to pay attention so we don't lose sight of that.

Pagitsas: What advice would you give the CEO or executive leadership team of a company just starting their sustainability journey?

Blum: Sustainability has to be part of your culture and your DNA. Again, this idea of adding a positive impact is central. Make sure you select leaders and promote people that will be best for this positive impact. I mention that because sustainability at Schneider Electric is not a one-man show led by myself or a very committed CEO. If you want sustainability to progress, it must be part of the culture, and therefore it must be embraced by all the company leaders and be sustained over time.

Pagitsas: As we come to the end of our discussion, tell me, what is your favorite book? Why is it you favorite?

Blum: My favorite book is *Born to Run: A Hidden Tribe, Superathletes, and the Greatest Race the World Has Never Seen* [by Christopher McDougall (Knopf Doubleday Publishing Group, 2011)]. *Born to Run* is a fascinating book about a guy intrigued by why people perform or don't perform in running. He did a lot of investigation everywhere in the world to understand why people have been successful in running or not.

I like it because I love running, which is part of my own well-being. When you get older, you always want to progress, and you realize that there are different ways to progress. In the book, you realize that you can progress in running by going back to the basics of nature, humanity, and sustainability. I found it quite inspiring. I wasn't expecting that message when I first read the book, and I love to read it again and again.

Pagitsas: Lastly, what drives you daily?

Blum: It's basically to work with people and to have a positive impact. Over my career, I've realized that the more you give to people, the more you receive from people. That's something I like to do every day. It gives me a lot of energy.

Scott Tew

Vice President, Sustainability & Managing Director, Center for Energy Efficiency and Sustainability
Trane Technologies

Scott Tew serves as vice president of sustainability and co-founder of the Center for Energy Efficiency and Sustainability (CEES) at Trane Technologies. Trane Technologies manufactures heating, ventilating, and air conditioning (HVAC) systems and building management systems with almost 40,000 employees globally and $12.5 billion in revenue as of year-end 2020. He is responsible for forward-looking sustainability initiatives to transition to more efficient and climate-friendly solutions and minimize resource use within company facilities. His recent leadership includes introducing the company's comprehensive 2030 sustainability targets and launching the largest customer-facing corporate commitment to combat climate change with the Gigaton Challenge.

Tew serves on the Advisory Council of the Corporate Eco Forum, as the board chair of the World Environment Center, and the chair of the US Business Council for Sustainable Development. In addition, he serves on the North Carolina Energy Policy Council. He was named North Carolina Industry Energy Leader of the Year by the

© Chrissa Pagitsas 2022
C. Pagitsas, Chief Sustainability Officers At Work,
https://doi.org/10.1007/978-1-4842-7866-6_16

Charlotte Business Journal. *He is the recipient of the "Leading the Way" award in CSR by* Corporate Responsibility Magazine.

Chrissa Pagitsas: What are the key pillars of Trane Technologies' sustainability strategy?

Scott Tew: We have three pillars—the Gigaton Challenge, Leading by Example, and Opportunity for All. With the Gigaton Challenge, we are committed to supporting our customers' reduction of their carbon emissions by one gigaton by 2030 with next-generation product and technology solutions. In Leading by Example, we target multiple activities within our own operations, including achieving carbon-neutral operations, a ten percent absolute energy reduction, zero waste sent to the landfill, and net positive water. Lastly, our third pillar is Opportunity for All, which focuses on creating gender parity in leadership and a diverse and inclusive workforce while investing in STEM [science, technology, engineering, and mathematics] education.

Pagitsas: What's the secret to success in driving such a large strategy for a global Fortune 500 company?

Tew: In my view, the secret to success for CSOs [chief sustainability officers] at large, global companies is to become good at finding ways to change major things often one small step at a time—a journey is accomplished with many milestones. As I often describe this to people, even to the sustainability team, this can be tedious, almost dull, work at times. Every now and again, you achieve a new rating or receive an accolade from an external group, and then people say, "Oh my gosh. What an exciting area you work in." And I think, "The work was not always exciting, but it is certainly rewarding to see the progress and the fundamental change that occurs."

When people ask me what I do, I respond, "I work to enhance processes by inserting sustainability-thinking inside existing tools that lead to changed behaviors and outcomes." As a sustainability leader, you're working with people across the company to help them challenge existing approaches to achieve a new outcome.

I recently was in a discussion about an internal carbon tax framework that we are considering for product-level or scope 3 carbon emissions. I worked with the internal team in thinking through a long list of questions, possibilities, and the implications for our product lines because any eventual sustainability solution has to fit within a business strategy. The goal was to help identify solutions that are a win for the customer and the business and that have solid sustainable outcomes. A lot of my day is spent in ideation and providing input on questions that can help land us in a different place, a better place, to be honest.

You also have to be able to see around corners. You have to be curious about emerging trends or those issues that are on the horizon and then ask, "How might this impact the company, and what actions should we take?" We not only need to understand these issues, but we have to find ways to take the "philosophical" out and determine how to provide context to internal teams, business managers, and enterprise functions in a way that leads to a path with improved outcomes and actions.

Pagitsas: Once you've seen an emerging trend or issue on the horizon, what is your approach to address it?

Tew: Ideally, we're adjusting existing processes and solutions as we go. The concept of sustainability integration is focused on adjusting the existing processes that are already working—akin to continuous improvement. A successful company—like Unilever, 3M, Emerson, Honeywell, you name it—has found a way to be successful within a repeatable process that provides unique value. The trick for a sustainability leader and team is to approach a successful process and say, "You know, if we tweak the process this way, we may find new, enhanced value that leads to more wins with customers or with how we operate a facility." The wins could be that the customers are more satisfied for new reasons beyond just cost and product performance.

Our salesforce, for instance, knows how to sell HVAC systems to customers—including those who own or manage very large, complex buildings. They don't need me to tell them how to do that. The trick is for me is to find out how to help a salesperson understand that the property they are calling on also has set a greenhouse gas reduction target. If the salesperson talks about our solutions and how they can help achieve the customer's greenhouse gas reduction target, you have enhanced the value proposition solving at least two customer concerns—comfort and environmental impact. The customer is happier because not only are you providing comfort for them through our equipment, but you're also providing a way for them to improve their own journey on greenhouse gas reductions. That dual win requires providing the salesperson a way to do something differently, such as setting a bold emissions reduction commitment that inspires action, adjusting the selling process, or possibly developing a tool that helps better explain or calculate the enhanced outcomes. That's the work that we do. Integrating sustainability is about adjusting existing processes.

Pagitsas: How has your senior leadership engaged on the topic of sustainability?

Tew: When Mike Lamach was named Chairman and CEO of the company in 2009, the board asked him about the possibility of using sustainability as a lever similar to lean manufacturing to help identify inefficiencies and as a tool for improving the business operating system. After that meeting, Mike and the leadership team, which includes Dave Regnery, the company's current Chairman and CEO, asked, "What if we integrated sustainability into the fabric of our businesses and functions?" In response, we developed the Center for Energy Efficiency and Sustainability to do the tedious, programmatic work of adjusting processes, providing training, and bringing new insights that led to a series of bold commitments that provided a compass and focus for our journey.

Pagitsas: How should companies like Trane in equipment manufacturing think about where to integrate sustainability into their businesses?

Tew: It's important to take a value stream approach. For original equipment manufacturers [OEMs], focusing on the product design process is critical. Product design and redesign typically leverage a consistent way in which the company develops, retrofits, and specifies the products or solutions, whether it's a tool, a software application, or a product category with similar characteristics. People who design products go into the design tools and develop a company's next-generation offerings. Engineers are knowledgeable in how to work within this design system. They design and refresh products when a business requests new features that customers want, for example.

In this case, it's important for the sustainability team to first understand the existing process and the existing tools and then find the right point within the process to help the engineers encounter and answer a new set of questions. For example, when you're specifying raw materials, is there a place in the design tool that provides a question that asks, "Should the product contain recycled metal?" To answer this, you will need a series of decision tree–type questions concerning the potential implications of recycled metal, like copper, versus virgin-mined copper.

But to accomplish all of this, you first must tweak the design system around a set of new, relevant questions. In addition, it may require new training for design engineers to help understand the context behind a new paradigm leading to enhanced value and the impacts of various scenarios.

Over time, you begin to fundamentally change future product categories. The result are next-generation products supersede regulations and that raise the bar for the category and at times for an industry.

Pagitsas: What is a misconception about sustainability that's important to dispel?

Tew: I think many people feel like sustainability at times can be restraining. The word *efficiency* makes them think, "I'm giving up something." Instead, sustainability is an enhancer, and it's about adding the word *and*. For example, you can say, "I want a beautiful, fast car *and* one that is highly efficient." Or, in our company's case, "I want an air conditioning system that provides comfort, *and* I want it to do in the most cost-effective and climate-friendly, efficient way possible." There's always an *and*. I think sustainability professionals should focus on the *yes, and*. We have to be good listeners with the ability to keep one foot in the present and the other in the future in order to come up with the *and* because so many good ideas have yet to be explored.

Pagitsas: Using the power of the *and* is a key element for success for a sustainability professional. What do you think is the fastest way for a sustainability professional to fail?

Tew: Not focusing on the highest impact area or not understanding the most non-financial material issues for their company. That's a setup for failure. It's important that sustainability leaders clarify and continue to refine the few high-priority issues—those that will make the biggest impact, that are the most meaningful, and that they become evangelists for those things.

Pagitsas: Let's talk about Trane Technologies' most material issues. What are they?

Tew: It's emissions reductions. Honestly, our overall greenhouse gas emissions footprint is too high. To address that, we have worked hard to understand the data. It's important to have the facts and the data to see the steps and develop the pathways required to make a difference. We need to understand the emissions profile and related issues so that the sustainability team can develop impactful plans and options to address them in the most meaningful ways.

Once you've identified the high-priority issues, it's also important to address the next steps. In my opinion, sustainability leaders who struggle do so because they have not yet identified the highest priorities or have yet to identify the steps necessary to make valuable improvements.

Pagitsas: Why is emissions reductions a material issue for Trane Technologies to solve with its sustainability strategy?

Tew: We provide cooling for the world, which is now warming, so it is a critical issue more than ever before. In a warming world, cooling has become a necessity. However, cooling and heating buildings via HVAC represents about forty percent of the electricity load, which equates to about fifteen percent of the world's greenhouse gas emissions. Trane Technologies also has a business responsible for cooling fresh food and perishables during transit to customers. In fact, we lead the market in ensuring perishables are effectively transported across highways, rails, and oceans. We cool it to keep it fresh and to prolong its useful shelf life. Food loss in the cold chain is responsible for about ten percent of the global greenhouse gas emissions. Let's add that up. Emissions related to cooling and heating buildings and in cold chain-related food loss equals about twenty-five percent of the global greenhouse gas emissions, which Trane Technologies is currently attempting to solve.

Are we part of the problem? Yes. Because our systems consume a lot of energy. Are we trying to solve the problem? Yes, with our innovations, research, critical thinking, and strategy. The company is focused on how we unwind from the current situation. That is our North Star. Thankfully, everyone inside the company gets it. Part of the job of our team is to identify options, drivers, and expectations from external stakeholders as we focus on solving these global problems.

Keeping the problems in focus, asking tough questions, and sharing insights drive collaboration and lead to possibilities. There is a deep passion behind it all. We've committed to reducing customer carbon footprint by one gigaton, which is really about changing our future product portfolio. It means understanding the issues well and then doing the hard work of bending the emissions curve for our customers and the planet.

Pagitsas: You've discussed addressing customers' emissions footprint. How are you addressing Trane Technologies' own emissions?

Tew: Within our operational footprint and supply chain, our goal is to achieve carbon neutrality. We're using science to drive decisions around emissions reductions, absolute energy reductions, and achieve net positive water, such as providing access to more water than we consume.

We've committed to partnering with suppliers in unique ways. If you are a supplier that has not set an energy goal, for example, or might not understand the options for reducing your energy profile, we will help you develop a plan, including providing a tool to help you track your progress. That's a true partnership. We're inviting our suppliers to join us in our journey, including in future renewable power purchase agreements.

We understand that we won't solve the problems alone. We need to collaborate and partner in ways that we've not done in the past. We've gotten good at partnering over the last decade, but like many companies, we have to do better at partnering with others in the future. And we may need to solve future problems in different ways like with crowdsourcing ideas or through gamification. We also realize that we're going to need to do it in a way that is inclusive of a diversity of people, views, and opinions.

Pagitsas: You've touched on customers and your operations. How are you engaging employees?

Tew: Employee engagement is critical to our efforts. As an example, we recently launched *Operation Possible*, where we invited our employees to give us their ideas about "absurdities." This initiative sends a clear message to employees: "We want your ideas—even your absurd ideas—that could help us solve a big challenge." If you have an idea that could help solve an absurdity and you have not found an outlet to discuss it, then Operation Possible provides you with an opportunity to engage. This is your moment. We've collected hundreds of exciting concepts and ideas that show real promise.

We're going to continue this with our employees, and we're exploring ways to broaden this outside the company. If there's someone out there who has invented a disruption that could lead to a next generation of air conditioning and wants to get in touch with us, this could be your moment.

Pagitsas: How has hiring for sustainability teams changed since you started in this field?

Tew: When I started in sustainability, I was looking for people who were technical yet were also good at ambiguity. There were very few credentials, and hardly any university had programs in sustainability. Today, I still look for people who are good at leading without authority and have unique skills. However, today the landscape of qualified candidates is immense. It is very common to find applicants with degrees in disciplines ranging from sustainable finance to sustainable design and with specialized credentials in green building sciences, ESG reporting, and issues such as recycling. In addition, many applicants bring several years of ESG-related experience with organizations working to solve complex issues like food loss. This is truly an exciting time for young professionals. The diversity of interest and the elevated focus within companies is a great indication of how this field is growing and how students are focused on applying these skills to affect positive outcomes.

Pagitsas: How has the COVID-19 pandemic changed the way that Trane Technologies supports its employees?

Tew: We learned a lot in the pandemic about our business, our commitments, and working differently in the future. Some of the issues were already part of our 2030 commitments, including the strategy pillar we refer to as Opportunity for All. It was based on the idea that there's a mega-trend around the future of work that is going to change the way we work. But we didn't really know the extent of what that meant when we set that commitment. COVID-19 provided a new lens for the future of work that has taught many lessons.

On the people side, our company has always been safety focused. We are an industry leader in putting employee safety first. We were working from home very early on in the pandemic. We also began to think through the flexibility needed to help employees continue to be productive at home. We introduced a new set of resources that supported working at home, including considerations for childcare and flex and hybrid work arrangements. Many teams have found ways to be highly productive in a hybrid fashion or a fully virtual fashion. It's clear that the future of work will be different for us all.

On the business side, our company has thought a lot about indoor air quality for many years. It's the bread and butter of what we do. In the past, though, the issue was not top of mind for building owners. However, the COVID-19 pandemic has certainly raised questions such as, "What does 'safe indoor air' mean in my home? At my children's school? And, in the office so that I can return?"

Our concern as a company was that we heard so many people in the media promoting filters as a quick fix. We were very concerned that schools and businesses would add a lot of things to the building that was window dressing or might come with terrible trade-offs. For example, you might increase your energy bill and lose all the efficiency gains of the building because adding new equipment without assessing the impact on the total building system can have unintended impacts.

So, in September 2020, we launched the Center for Healthy and Efficient Spaces [CHES]. We are working to connect this idea of healthy buildings with efficient buildings in order to minimize the negative trade-offs. We believe that you can have a healthy building, and it can still be an efficient building. In addition, we created an outside advisory council of some of the world's leading scientists and building experts to advise us in this journey.

This concept is important because people often think that there are no negative trade-offs to doing things related to efficiency or sustainability. People may say, "We want healthy at any cost." Our response is, "You can have healthy without necessarily increasing costs." This *yes, and* approach is a unique stance in the market.

I think most companies, including Trane Technologies, have come away from the pandemic with some deep learnings of people, safety, expectations, working together, teaming, partnering, and what it takes to solve your customers' big issues in particular.

Pagitsas: If there is one message that you would give to the non-sustainability business leader, what would that be?

Tew: My message is that sustainability adds value in new and exciting ways. A better way is possible!

Frank O'Brien-Bernini

Senior Vice President, Chief Sustainability Officer
Owens Corning

Frank O'Brien-Bernini is senior vice president and chief sustainability officer (CSO) at Owens Corning. Owens Corning is a global building and construction materials leader with 19,000 employees in 33 countries and $7.1 billion in revenues in 2020. Frank holds global accountability for Owens Corning's sustainability strategy development and implementation, including operations sustainability; environmental, health, and safety (EH&S); product and supply chain sustainability; corporate medical and wellness; product stewardship and toxicology; and the company's science and technology center in Granville, Ohio.

Since joining Owens Corning in 1983, Frank has held multiple leadership positions, including chief research and development officer, before becoming chief sustainability officer in 2007. Before joining Owens Corning, Frank ran a solar design/build firm.

Frank leads routine strategy, execution, and impact discussions with the Owens Corning board of directors. He is engaged in new board member onboarding and has held multiple board positions, including for Powering Ohio, RESNET (Residential Energy Services Network), the Vytec Corporation, the NAHB Leading Suppliers

© Chrissa Pagitsas 2022
C. Pagitsas, *Chief Sustainability Officers At Work*,
https://doi.org/10.1007/978-1-4842-7866-6_17

Council, the Ohio Home Builders Association, the Center for Multifunctional Polymer Nanomaterials and Devices, and the Ohio Biopolymers Innovation Center. As a RESNET board member, he envisioned and founded their Supplier Advisory Board to add a supplier's perspective and significant funding to RESNET.

Chrissa Pagitsas: As of 2021, you've been at Owens Corning for over thirty-eight years. Could you take us on a journey of your career and how you became Owens Corning's first chief sustainability officer?

Frank O'Brien-Bernini: It is a long journey. It perhaps started when I was a child, as my mother was an environmental advocate. We lived in a neighborhood with a chemical manufacturing plant in our backyard, so she became a very strong advocate for pollution reduction. That was maybe my introduction to environmental advocacy. As a child, I observed the complexity and challenges of businesses impacting the environment and the surrounding community. Many of our neighbors worked at the plant, so they weren't exactly happy with her advocacy around what might end up causing a loss of jobs at the plant.

When I was in college, I had an opportunity to be in the first cohort of a new discipline at The Center for Resourceful Living at North Adams State College, now called the Massachusetts College of Liberal Arts. The Center for Resourceful Living was a mixed discipline program between the physics department and the history department, blending the natural and social sciences, focused on what we now call *sustainability*. It was focused on renewable energy, organic agriculture, sustainable farming, all that sort of thing. As a result of the program, I got very interested in solar energy, energy efficiency, and other types of renewable energy, quit school, and started up a design-build firm for super-insulated passive solar houses in southern Vermont.

I ran that business for three years before deciding I really liked the engineering part. I went back to school for mechanical engineering and focused on solar energy. I got my master's in mechanical engineering and wrote my thesis on solar energy. Owens Corning supported my work through a fellowship, and that's how I got introduced to the company. After I presented my fellowship work to Owens Corning thirty-eight years ago, they offered me a job.

Through the course of my career at Owens Corning, I have held various research and development [R&D] positions, all the way to my last role as the chief research and development officer, where I began advocacy for a sustainability strategy. In 2002, the CEO at the time said to me, "That's cool. I don't understand what you're talking about, but that's great if you can do that and your research and development job. Let's do both."

So, that's when I first became a combined chief R&D officer and chief sustainability officer. Then in 2007, the incoming CEO asked me to focus full-time on sustainability. So, it's been my full-time focus since then. That's my semi-short story.

Pagitsas: You're not a Johnny-come-lately to sustainability.

O'Brien-Bernini: Yes, that's for sure.

Pagitsas: Then let's do a compare-and-contrast exercise on what sustainability meant for Owens Corning in 2007 when you stepped into that new leadership role and what it means today in 2021.

O'Brien-Bernini: It's been a fun progression. I think Owens Corning's progression in sustainability mirrors the growing breadth with which the world considers sustainability. When we started, we were among the first companies to establish what I would call "holistic footprint reduction goals." That's how I think most companies have started out—trying to get their own house in order.

When we got started in the early 2000s, the IPCC [Intergovernmental Panel on Climate Change] had just issued a report in April 2002 underscoring the link of human activity to climate change. I think that was a wake-up call for many industrials that manufactured energy-intense, carbon dioxide-intense products. We began trying to get our arms around our energy use and carbon dioxide emissions. Almost concurrently with that, the US Green Building Council and other green building programs were beginning to gain traction with their certification point-structures and programs for green buildings, such as Leadership in Energy and Environmental Design [LEED]. Our business is focused on the building materials space. So, while it was nascent at the time, it became clear that there was going to be a growing focus on green buildings and a market opportunity for those of us manufacturing the building materials.

We began our sustainability journey thinking in terms of shrinking our environmental footprint and increasing our product handprint. In essence, doing less bad through environmental footprint reduction and doing more good through more sustainable products increases our handprint. That's where I think we and most of the world started this sustainability journey.

Over time, our sustainability journey got more expansive. We included our community outreach and supplier expectations. We then embarked on a wellness journey to eliminate lifestyle-induced disease and a safety journey to eliminate injuries at work and at home. We are continually growing the expectations and the ambitions of our sustainability strategy, including introducing inclusion and diversity into our sustainability strategy.

Most recently, we made clear our aspirations around circular economy and the full life cycle of our products, not just the parts which Owens Corning controls. Over time, the world and Owens Corning have increased the depth and the breadth of our ambitions. Concurrently, what has changed most recently is that the expectations of ESG [environmental, social, and governance] investors have gotten much more granular.

We're seeing now with the environmentally or "E-focused" investors and customers that the conversations have gotten much more specific about what the product line opportunities are for Owens Corning in a decarbonizing world. They want to know what solutions we have that will create growth opportunities for our company and help our customers grow. It's progressed from a more general conversation of, "We want to make sure you guys are 'contemporary' in this ESG space," to "What are you specifically going to bring to market?" and "What's the embodied carbon associated with your product lines?"

Pagitsas: How do you decide the top issues to tackle when you have many investors and customers with different priorities and strategies?

O'Brien-Bernini: Through a materiality assessment.

Pagitsas: How do you define a materiality assessment?

O'Brien-Bernini: At the highest level, it's an analytical assessment to figure out what matters most. Just like with financial materiality, it's what matters most to the stakeholders that are invested in some way in your company. Those can be employees, the community where you operate, your customers, and certainly investors. It's an active assessment to figure out their expectations are of your company. Then, you can figure out where you should focus your resources and set goals.

Pagitsas: On the latest 2019 materiality assessment for Owens Corning, why was circular economy one of the highest-scoring material issues for your internal stakeholders as well as your external stakeholders?

O'Brien-Bernini: I think that the world is realizing that climate action is the North Star. There are a lot of conversations around a circular economy. However, I think of this as subservient to the North Star of climate action and reducing carbon dioxide emissions. For us, and I generally think for the world, a circular economy is a powerful tactic in the pie chart of options, technologies, and approaches to reducing climate impact.

A circular economy is one in which virgin raw materials, waste, energy, and emissions are minimized through intelligent design, renewable and recycled inputs, energy-efficient production, and enabling the recyclability of products at the end of their life cycles. Our aspiration is that all materials extracted to make Owens Corning products remain in the economy indefinitely. For our business, this is a powerful climate action lever. For example, when we use more recycled materials to make our products, it reduces the embodied carbon or the carbon dioxide emissions associated with manufacturing those products because it takes less energy to manufacture with material that's already been processed. If we use waste glass to make our new glass, it's less energy-intense and less toxic air emissions intense than using basic raw materials like sand, alumina, and borate.

We see the circular economy as a "megatactic" in our progression from trying to reduce the energy that we use by getting more energy efficient and then making large purchases of renewable energy. As we develop along this waterfall from reducing carbon dioxide emissions to total decarbonization, the circular economy is a significant tactic to help us attain decarbonization.

Pagitsas: Is there a particular product that Owens Corning manufactures that you feel has attained the highest level of carbon reductions and is representative of a truly circular economy?

O'Brien-Bernini: We have no products at the highest level because we have high ambitions. However, we do have products that are very far along the path. Today, we use roughly seventy-percent recycled content in our fiberglass insulation products. We certify that recycled content is between fifty-three percent and seventy-three percent, depending on the specific product. That recycled glass is largely from beer and wine bottles, plate glass from manufacturing and building deconstruction, and the manufacturing and end-of-life operations of window glass for automobiles. We are very heavy into the use of recycled materials today. What that does is eliminate the need for mining of sand and borates, alumina, limestone, and those kinds of typical raw materials used to make glass, so you get the energy and embodied carbon reduction associated with that averted raw material extraction.

We also get a large reduction in energy use in our furnace operations when we melt glass instead of melting virgin raw materials. We have fewer emissions from those furnaces because remelting glass has a lower emissions profile than melting virgin raw materials. Lastly, we can market the product as a high recycled content product.

Pagitsas: There's clearly an environmental benefit and a business case to use recycled material. Are there community or social benefits?

O'Brien-Bernini: Yes. There's a large social benefit to community recycling. Local job creation occurs in the recycling processes, whether through hiring people to do curbside recycling or to work at MRFs [municipal recycling facilities.] Job creation is also associated with the recovery of those materials that tend to be higher per ton of usage than mining extraction, which tends to be very machine intense. There is also local job opportunity. The materials we use, sand, limestone, and those kinds of things tend to be available very locally, so we do, in general, source our materials very locally. The recycling value chain has more job opportunities than virgin raw materials generally.

Pagitsas: Coming back to the investor perspective, what is the conversation you're having today with investors?

O'Brien-Bernini: Yes. Right now, there's a segmentation of investors into three categories. They may segment themselves somewhat differently, but from my perspective, we have investors who are more "traditional investors" and the conversation is dominated by finance, revenue, forward-backward projections, and markets, and followed by a relatively minor box to check around making sure that our company isn't doing anything that they, as an investor, would be embarrassed by or would generate an unexpected environmental, social, or governance risk.

However, I think that's a dwindling segment. Investors are generally moving into the next two boxes. The next type of investor has a high interest in ESG, where they want to really understand on the *E* or environmental side, what are we doing around footprint reduction and decarbonization. And, on the *S* or social side, they want to understand our inclusion and diversity, board structure, what are we doing around attracting more women, and how we are increasing the representation of racially diverse minorities in leadership and on the board. They are very interested in all segments of ESG.

The third segment of investors may be niche, but it's a growing niche. There is a fair amount of our investors now E-focused, and they are building a portfolio of companies that will allow them to position their fund with their investors as a climate action fund. They want to know what we are doing to advance energy efficiency in buildings globally, the European Union, and in the United States, specifically given our building materials exposure in those two regions. They are very interested in the growth opportunity and total shareholder return opportunity in our stock relative to the positive exposure we have regarding what they care most about as they build their investments for their climate action fund. Those are the three areas where we have most conversations with our investors.

Pagitsas: Suppose an executive said, "I don't want money from that third segment of investors because their climate action requirements will be more than I can handle and will open me up to liability." How would you respond?

O'Brien-Bernini: That's a great question, and in some ways, to be totally respectful, an irrelevant one. [laughter]

Pagitsas: Great, I'll play the devil's advocate. Why is it irrelevant?

O'Brien-Bernini: What the investor is trying to understand is our business strategy. We should never shy away from our business strategy. For example, one aspect of our business strategy is to grow our impact in the wind business. We supply glass fibers and fabrics that go into those huge wind blades that enable larger wind turbines. Our materials make wind blades lighter, longer, stronger. This means that our customers, the big wind OEMs [original equipment manufacturers] who make the turbines and the wind blades, can make larger and larger turbines for onshore and offshore applications.

When the investors are trying to understand our positive exposure, they want to know what we are doing to secure our place as important to the growing wind industry. If they're interested in the building material space, they want to know what we are doing to develop products that are selected in the green building and high-efficiency building space.

We're not trying to figure out what goal we should set for the company so that we're attractive to an E investor or an ESG investor. We're trying to transparently communicate our business strategy, and if it aligns with their investment strategy, they're a great partner or a great owner of our business. We're not trying to attract investors by setting ambitious goals that attract them. We're trying to communicate our business strategy.

Pagitsas: Let's say that the executive still responds, "Thanks, Frank, but I'm doing fine. My core strategy is fine and profitable without reducing its carbon emissions. I'm not going to integrate sustainability into my core business." What would you tell them?

O'Brien-Bernini: I'd tell you that you're missing an opportunity. I'm not saying you're wrong, but what I'm saying is that the world is committed to decarbonization. You may be doing just fine, but you are going to have a growing number of customers who ask you generally about your sustainability strategy, or specifically about your carbon emission reduction targets because you're a supplier to them and what you do or don't do will impact their ability to meet their sustainability goals.

There is a growing number of customers who five years ago had a conversation with us about our goals, a very cordial conversation, and asked, "What are your commitments? What are you focused on globally as a company?" It was a very nice, pleasant conversation, and everybody was happy to do business with each other.

It has gotten way more granular in the last year or two. Some of our largest customers and some of our smaller, fast-growing customers want to specifically know the carbon dioxide emissions associated with the products that we sell them. Not of the company, not about a global goal related to carbon, but the specific materials made in our plants, transported in this particular way, that arrive at their dock. They want to know what the embodied carbon of those products is.

If you, as the executive of the company you described, are not able to answer that question because you've said it's irrelevant your product performs at the right price today, my view is that you're going to be "despecified," or in other words, fired as a supplier. A customer may specifically state "not XYZ company's product" because they don't know its embodied carbon. That's a problem for that supplier. We never want to be that XYZ company!

Pagitsas: You're saying that architects are explicitly telling their customers which products to buy and which not to buy, depending on whether the manufacturer discloses the product's embodied carbon.

O'Brien-Bernini: Yes. That's what I see as the risk in that strategy. You may say that you don't believe in the importance of carbon emissions reductions. And it's fine that you get to run that company as you wish. But if you're trying to attract investors, like the E investors or the ESG investors I was talking about, you won't have them because they're going to say, "These guys are clueless." At a minimum, they are going to say, "They're not positioning themselves to take advantage of an obvious secular trend that we see playing out over the next few decades, as the world aims to decarbonize by 2050 at the latest. I believe the world is growing in its resolve to decarbonize, consistent with the IPCC's 1.5-degree-Celsius maximum global warming pathway.

Pagitsas: You're clearly laying out an argument that a company that does not focus on reducing emissions is missing out on the direction investors and customers are moving in. Taking it back to the product level, why are your customers asking you for the carbon footprint of the products that Owens Corning produces for them?

O'Brien-Bernini: They are responding in many cases to their customer. We are one, two, or three tiers back in some of the supply chains. We talked earlier about the wind business. We supply the glass fibers that go into the wind turbine blades. They get attached to the turbines that then get sold to wind developers. And those wind developers either sell the developed project or sell the power.

Somewhere in that value chain of decision-makers, someone is asking, "What is the embodied carbon of your turbine?" That's because the decision-maker is trying to convince a government, for example, that wind power is a better option than natural gas. To have that science-based conversation, decision-makers need to know the full life cycle impact of the wind turbine operating for twenty years versus a gas turbine that's also operating for those twenty years. Therefore, they need to go back in the value chain to us, at Owens Corning, who manufactured the glass fiber, and find out the carbon emissions. Because for a wind turbine, virtually all the carbon emissions are upfront.

After you build it, wind is essentially carbon-free for twenty years. Largely there's no carbon impact, not zero but near zero for its operations, but it's not near zero to build. If you're going to have that science-based conversation with the government, you better know upfront what that embodied carbon is. They need to know from us what the carbon is associated with our glass fibers that go into the wind turbine blades.

It's all in response to the customer's demand to know the impact so they can effectively represent their product. It's the same thing for automotive. Automotive is rolling out EVs [electric vehicles]. There's a lot of conversation around the embodied carbon associated with manufacturing EVs, and there are, of course, glass fiber composites in those EVs.

The automotive OEMs then ask our customers, who are the thermoplastic compounders, "What is the embodied carbon of the materials that are going into this EV that is lightweighting it?" Then we can have an EV life cycle conversation versus internal-combustion-based vehicles that may be heavier and use more steel. It's a business question that's being asked, not a sustainability one, but it takes deep sustainability skills and capabilities to answer.

Pagitsas: We've discussed Owens Corning's customers from a business-to-business context. Is there feedback, influence, or discussions with individual consumers about Owens Corning's sustainability strategy or the sustainability of its products?

O'Brien-Bernini: Because our business is virtually all business-to-business, our consumer interactions are secondhand or thirdhand, informed by what our direct customers are trying to accomplish with their customers. That said, our employees often provide feedback on our approach, and sustainability is an increasingly important conversation for recruiting. It's no surprise that individuals who work with or consider working with Owens Corning pay attention to the "say/do" of our sustainability work. People want to make a difference, and they want their work to be aligned with their personal values. And, of course, employees are also consumers—many of them have Owens Corning products in their homes and all around them. They are an important voice in ensuring that we take our commitments seriously and focus on the things that matter, which is an important input to our routine materiality assessments.

Pagitsas: You've been at Owens Corning for thirty-eight years, so you are very familiar with the business strategy planning process. How do you plan at Owens Corning? Where does the sustainability strategy fit into the planning process?

O'Brien-Bernini: We have a running three-year long-range planning process where we set the business strategy over a three-year timeframe, and then year one of that business strategy is our operations plan. Within that, there's the expectation that the business strategy and the sustainability strategy are totally aligned through that three-year timeframe.

The way that it all works, and this is what I love most about sustainability in any company but certainly in ours, is that it's the only area in our company where we set very long goals. For example, we need to roadmap a ten-year goal for a fifty-percent reduction in carbon dioxide emissions. Therefore, the core function of my small sustainability team is to make sure that in our business, long-range plans—which are three years, not ten years—include the very granular projects to make progress toward the ten-year goals.

Pagitsas: Today, Owens Corning as a company is organized into product verticals. One vertical is roofing, another is composites, and the third is

insulation. What falls within your team's responsibility, and what falls within
the business product verticals to execute the strategy?

O'Brien-Bernini: Largely, our sustainability projects and initiatives are
executed in the business, where the bulk of the resources are, not by our
core sustainability team. While we provide the roadmap, the waterfall of
options to attain the goals and the talent and financial resourcing comes from
the business vertical. For example, we have a fifty-percent greenhouse gas
reduction goal. That's a science-based target. We have a waterfall of tactics to
attain that goal, which includes using different material systems that emit less
in processing. Another example may be increasing the use of renewable
energy, from where we are today at around fifty percent, up to 100 percent
by 2030. Our work is to make sure that the projects exist in the business
plans. We think about it as moving from presentations to projects. The
projects existing in the businesses are going to allow us to attain our
enterprise goals.

What we bring to the table is technical expertise, such as understanding how
our products can contribute to a building's LEED certification points. However,
our job is also to make it clear to the businesses where the puck is moving on
the ice, such as the market's growing focus on GREENGUARD certification,
which addresses emissions from a product and is important for good indoor
air quality. We always want to share an informed perspective that, three years
from now, this red list material that you've got in your product is not going to
be acceptable because of either LEED's or another certification's new
requirements. Our job is to see further down the playing field and make sure
that the projects exist and deliver on the business we want to be operating
one, two, three years from now—and one, two, three decades from now.

Pagitsas: Does your team own the project? Or does the business leader,
such as Marcio Sandri, who is the head of the composites business?

O'Brien-Bernini: Yes, to both! Maybe two different stories that have two
different angles will help illustrate how we structure project ownership.
Related to composite glass fiber, fabrics, and nonwoven mat materials
manufactured all around the world, we have a 2030 goal to eliminate waste
coming out of our manufacturing operations and going to landfill. Glass fiber
waste is hard to recycle. It's a real challenge. While we want it to go away and
we've got that ambition, we have to have projects that solve the technical
challenges, like the engineering challenges of affordably recycling glass fiber
back in our composites glass furnaces. Same thing in insulation, but let's stick
with composites.

In real time, I have a director of circular economy manufacturing solutions
who ensures that projects related to the glass fiber recycling goal are being
resourced with the right people, figures out the technology needed to solve
the challenge, and identifies how it will be funded with capital to execute the

solution. Teresa Wagner is the director. She has accountability to ensure that there's a project portfolio to attain those goals. She works collaboratively with Marcio's team and Umberto Rigamonti, the VP of operations. Teresa and Umberto are connected at the hip with a joint goal to eliminate waste to landfills for composites. They have pilot projects to test technologies to make sure that they work.

It's a collaboration, but Teresa is the person who says if we are on track or not. She has accountability to ensure that we meet our 2030 goal and yell like hell if her portfolio doesn't roll up to that. The action that would occur from yelling like hell is adding resources from the business to the project to ensure the goal is on track to be attained. It is very much a team sport.

The other example I wanted to give is that not all projects work that way. We aspire to operate with 100 percent renewable electricity. To achieve this goal, there is a small team working on this among my team, the finance team, and the sourcing team, which does all energy sourcing. They source fossil fuels as well as renewables.

That three-legged team between finance, sourcing, and sustainability is the one that goes out and puts out the RFPs [request for proposals] and closes the deals to source renewable energy for our company. This type of project, where the beneficiary is the entire enterprise, isn't completed in partnership with the three business units. They're a recipient of that great work.

Pagitsas: Is there a story that captures a memorable success in working toward a goal?

O'Brien-Bernini: We've talked about the power of long-term goals and strategic ambition. The story of our recent product launch, Foamular NGX, is a great example of this. As I mentioned before, regulation is often slower than customers when it comes to demanding sustainable products, so going beyond compliance is part of our business ambition. We have been working to reduce the climate impact of our extruded polystyrene insulation product since the early 2000s, striving to stay ahead of regulations while meeting our customers' needs.

While we have made steady progress on this, our new Foamular NGX is a huge step forward, delivering an eighty percent reduction in embodied carbon without compromising performance. Over two decades of development, we have had six different general managers running this business. However, our clear ambitions and goals have kept this very complex research and development work on track to deliver this important step forward. This, more than any other story I could share, exemplifies the power of ambitions, long-term goals, and an aligned North Star!

Pagitsas: Has there been an equally memorable challenge or barrier that was overcome or resulted in significant "lessons learned"?

O'Brien-Bernini: When we first embarked on our work to develop a formaldehyde-free fiberglass product for the residential insulation market, we were very focused on this single "sustainability attribute." As we engaged our customers with our in-progress prototypes, it became clear that, in addition to delivering a new greener attribute, the product's core function needed to excel in areas such as installation speed and performance.

Many customers used the CFL [compact fluorescent lamp] lighting analogy, saying that they would not compromise core functionality to gain even an important sustainability attribute like formaldehyde-free. This was at a time, before LEDs [light-emitting diode], where CFLs were very energy efficient yet had a long warm-up time before gaining full illumination. We adopted a phrase that we still use today in our development work which is, "green without compromise." This continues to serve us well in framing "what is winning" in our new product development.

Pagitsas: I am frequently asked to advise where a company should place their sustainability team within their organization. Where do you think a sustainability team should be?

O'Brien-Bernini: Everywhere.

Pagitsas: Agreed! The sustainability team needs to partner with every part of a business. From an organizational chart perspective, who do you report to?

O'Brien-Bernini: I report to the CEO, and then my organization reports up through my leadership team. We're a fairly flat organization. My team's work is connected directly to the CEO. My direct reports are responsible for environmental, health, safety, and sustainability, and there's one layer of individual contributors and project leaders that report to them.

Pagitsas: You're really at center court.

O'Brien-Bernini: Exactly, yes.

Pagitsas: Let's talk about social issues. How has your company's strategy changed to address diversity and inclusion within Owens Corning?

O'Brien-Bernini: The "change" is just an acceleration of the work we'd already started. This is another example of how our long-range sustainability goals help us get and stay on the right path. In 2019, our new CEO, Brian Chambers, created a new VP of inclusion and diversity role. As we set our 2030 sustainability goals, there was massive executive support for having a formal inclusion and diversity goal. We'd been doing the work internally, but this was the first time we set broad, public, long-term goals. Because of those goals, we'd already begun having difficult, important conversations about these topics.

So, for example, when the critical conversation about racial and social injustice ignited in the United States via the Black Lives Matter movement, we already had a foundation for what we needed to do. But it became immediately clear that we needed to do more and do it faster. We increased the pace of our internal programs and training and started looking for ways to engage with our community. In line with our goals and purpose and employee and community expectations, the company—and Brian himself—wanted to be more vocal externally. From making statements on social media to drawing together CEOs and leaders in the community around our global headquarters in Toledo, Ohio, and committing over $6 million to advance racial equality and social justice in Toledo, we've been working on adding our voice and our resources to making a difference.

Pagitsas: What are you hopeful about as it relates to the intersection of sustainability and business? What are you worried about?

O'Brien-Bernini: Two great questions. I would say that five years ago, I was worried that there wasn't enough market pull to drive sustainability solutions at scale. I am not worried about that anymore because the investor questions in this space are getting more pointed, more granular, and less polite, which for me, is helpful. There's nothing more helpful to move a strategy forward than investors, customers, and employees directly engaging our executives or board on these topics.

Maybe that's what I used to be worried about, but I'm not anymore. That makes me very hopeful that we will use market pull to drive climate action. Expectations around embodied carbon, decarbonization, and so forth are going to move the needle from watching regulation to watching customers, which I think is way more urgent and compelling for most organizations because people often take solace in how slowly regulations move, but customers can move like that. [snapping fingers] They can just decide, "We're not happy with you anymore because we don't think you're moving in alignment with our strategy." I'm very bullish and hopeful that climate action will be moved forward in a big way because of market and investor expectations and pull.

What am I most worried about now? I think everybody is worried about speed—the speed of impact on how to be bigger and move faster and further. This has got to be on everybody's mind, whether it's about social justice, environmental justice climate action. Everyone is getting more impatient to move faster, which is important.

Pagitsas: What drives you daily?

O'Brien-Bernini: The opportunity to drive impact at scale is a huge benefit in working to advance sustainability in a large, for-profit company with the resolve and resources to make rapid change.

Pagitsas: What book inspires you and why?

O'Brien-Bernini: At my core, I'm a research and development, practical-innovator kind of guy, so I like the book *Drawdown: The Most Comprehensive Plan Ever Proposed to Reverse Global Warming* [by Paul Hawken, editor (Penguin Books, 2017)]. The book breaks down the huge challenge of climate change into pie segments of actions that can be taken to mitigate climate change. It turns a very thorny, amorphous challenge of decarbonizing the world into the potential impact of buildings, the industrial use of energy and electricity, and transportation. Inside of that, it asks what are the large things that can be done at scale to impact that segment of the pie chart.

I love the book because it aligns very well with the way I think about sustainability in our company. It frames the North Star, and what we are trying to do with our purpose, people and products make the world a better place. It helps me ask my team, "How do you get from that huge, purposeful ambition all the way to the project you are working on today?" That's what I love about that book. I refer to it often because it's got a lot of very useful data.

The people that are behind that have continued to do that work, and the book includes all those citations so you can follow the ongoing work they're advancing through technical journals. I really like that. Maybe that's a boring answer because it's my job, and the good thing is that my definitely not-boring job is super-integrated into my life.

Pagitsas: Who or what experience has most influenced your leadership style?

O'Brien-Bernini: There are four competencies that I find critical to the CSO role, and for me, each has a history of influence on me. These competencies are change leadership, being driven by purpose, harnessing the power of storytelling, and integrating work and life.

Leading change successfully is at the heart of a CSO's role. I am a fan of both Simon Sinek's why-how-what model and John Kotter's eight-step model to successfully drive change. Simon Sinek's model forces you to put things in the right order to engage others, not surprisingly, starting with why your company or initiative exists. John Kotter's model helps make sure you have thought through and make good use of all the proven steps to effectively drive change. These are very solid tactics that I draw on often.

Related to being driven by purpose, I was greatly influenced by Larry Vadnais, my professor and the force behind The Center for Resourceful Living at North Adams State College. He helped me see the value of declaring purpose and aligning to that North Star every day as you make large and small decisions.

Our former CEO, Dave Brown, was a master of storytelling, and I learned a lot from him. Once we were preparing a board presentation on the handprint or positive impact of our products, and I recall focusing on the engineering numbers. Instead, he helped me construct a powerful story, converting the numbers to a more understandable analogy. The story I ended up sharing with the board was, "Let's look at the handprint or footprint 'math' of a typical home built in Chicago. With the energy it takes to manufacture, transport, and install our EcoTouch insulation to meet the energy code in this 2,400 square foot Chicago home, you could drive a typical car across the United States about 3,000 miles. That's a lot of energy. Now let's look at the energy saved by that insulation over a modest sixty-year lifetime. Well, you could drive that same car to the moon and back about a 478-thousand-mile round trip, not once, not twice, but five times! That's the math of 'doing more good.'"

Lastly, from my own experience, work/life *integration* is critically important to personal and professional happiness. This is distinct from an aspiration of work/life *balance*. This pursuit of integration, a realization I have come to through my career, helps me relieve the normal stresses associated with pursuing work/life balance by, instead, seeking to totally integrate my work life and my personal life. Given my personal convictions and the role of CSO, this has been a welcomed progression! My complete, all-in life is everything I love.

Financial Services, Investment Management, and Commercial Real Estate

Elsa Palanza

Managing Director, Global Head of Sustainability & ESG
Barclays

Elsa Palanza is managing director and global head of sustainability and ESG [environmental, social, governance] for Barclays. Barclays is a global bank with 83,000 employees and more than £1.3 trillion in assets as of 2020. Elsa joined Barclays in 2018 with over 20 years of experience as a strategic advisor for corporations, foundations, and non-profits.

In her role, Elsa leads the strategic direction of the bank's policies and practices across a broad range of sustainability and ESG issues, including climate change, environmental stewardship, human rights, and social impact. She heads external engagement and partnership development on ESG matters, engaging with clients, investors, civil society groups, and other diverse stakeholders. In addition, Elsa has overseen the development of standards and metrics to advance green and sustainable finance and steward early innovation in sustainable product and service development.

Prior to her role at Barclays, Elsa served as director of commitments for the Clinton Global Initiative. She also created and launched a platform to advance progress on the UN Sustainable Development Goals (SDGs) for the Bill & Melinda Gates Foundation. Earlier roles include delivering geopolitical and industry risk analysis to

© Chrissa Pagitsas 2022
C. Pagitsas, *Chief Sustainability Officers At Work*,
https://doi.org/10.1007/978-1-4842-7866-6_18

large international energy companies and leading strategy and business operations for an education company.

Elsa serves on the Advisory Boards of First Book, a non-profit social enterprise focused on education equity for kids in need, and Masawa, a mental wellness impact fund. Elsa appeared in the award-winning film, Our Planet: Too Big To Fail, *which explores the role the finance sector can play in powering a sustainable future.*

Chrissa Pagitsas: Your background in social impact, energy, and geopolitical risk analysis is unique for an executive in the financial services industry. Describe your professional journey to Barclays and this role.

Elsa Palanza: I've had an entirely nonlinear career path, and for that, I am extremely grateful. Now that I am further into my career, it has ended up serving me very well. My career has always been driven by a sincere sense of wanting to do what felt in service to others and authentic to who I am.

When I graduated from college with a degree in international relations, I moved to Washington, DC, and slept on a friend's couch while I searched for a job. Being a girl who grew up in a family of artists in New Mexico, I didn't have any language to describe what I thought I wanted to do. But I knew I wanted it to involve international policymaking. I ended up getting a role in a boutique consulting firm that specialized in energy policy and geopolitical consulting.

I had gone to DC because it felt like that was the heart of where things were happening, at least at a policy level. Indeed, it was important to learn how the interagency process works—or sometimes doesn't work—and understand why the complex negotiations and conversations that go on behind closed doors between government and the private sector are so important. They often show up later down the road in ways that you might not have anticipated.

Our firm's clients were some of the world's largest energy and utility companies. Getting to know the humans behind the companies and understanding what they were grappling with within the energy industry was not only tremendously interesting back then, but fast forward all these years to now, it has been an important feeder for my understanding of the factors impacting the transition to a low-carbon economy for the energy and power industries. It's a whole system change that must happen.

Pagitsas: What key experience from this first professional role influences you still today?

Palanza: I learned that partnerships across sectors and political lines can and must work. My boss at the time, who was the chairman of the company and to whom I reported directly, is an extraordinary thinker. He and I had very different political persuasions, but he was willing and interested in making sure there were partnerships and opportunities for collaboration at every juncture of the road. He had been in the Reagan Administration but had incredible super lefty friends from the United Nations. You'd never think that these characters would be friends and collaborators, but they were. They respected each other and leaned on each other to get better intelligence about market events and policy ideas. That theme of partnering across perceived boundaries has come back around for me many times. This experience was an important foundational chapter in my career.

Pagitsas: What was the next chapter of your career?

Palanza: I moved to Istanbul, Turkey. It was less about my specific job there, frankly, and more about trying to get something done as a twenty-five-year-old woman in a country where I didn't speak the language and where cultural and professional norms were very foreign to me. It was hugely adventurous and wonderful. After a year in Turkey, setting up a language program supporting business and law students, I moved back to the United States and studied international relations at The Fletcher School of Law and Diplomacy at Tufts University. It's an interdisciplinary program where I took classes in everything from water and diplomacy to international security studies.

That community led me to the next important chapter of my career, which was moving to New York to work for the Clinton Global Initiative [CGI]. President Clinton is an extraordinary leader on several levels, and everything everyone says about his charisma and his ability to draw people in is true. His genius is really apparent in his ability to bring seemingly disparate pieces of information together. When he took a matrixed, complex issue and synthesized it down to the core of what is at stake, he crystallized for others the difficult decisions that needed to be made or the difficult relationships that needed to be unpicked or drawn together.

I can only ever hope to emulate some of that, but I think watching it in practice—both from the President as well as the incredible members we worked with every day—was incredibly impactful professionally. At CGI, our whole reason for existing as an organization was to build partnerships across sectors in order to solve big global challenges. Anything that could fall within the ESG spectrum—from social inequality to energy to climate change to green infrastructure and the financing that underpins it all—was all part of the conversations that were happening at CGI.

A lot of people knew CGI for the fanfare and the star-studded part of it because you did have big-name people showing up for events. While that was exciting, when it came down to it, it was what they were there to do that was most important.

Pagitsas: What was your role at CGI? Was there a key takeaway from your work?

Palanza: I led the team of subject matter specialists who helped CGI members develop what we called "Commitments to Action" or new, specific, and measurable projects to solve particular global challenges. My team was made up of specialists in different development areas, who worked on everything from big picture issues, such as enabling effective impact investing, advancing behavior change in global health, and increasing quality and access in global education, to very specific interventions such as growing employment opportunities for refugees, strengthening the supply chain for sustainable palm oil and cocoa, combatting elephant poaching, and much more. We worked with our member organizations—a broad community of companies, foundations, NGOs [non-governmental organizations], governmental and multilateral agencies, and other influential leaders—to determine where they were best positioned to make a positive impact on a particular social or environmental issue, create a time-bound action plan and the resources needed to achieve that plan, and articulate the metrics they would use to measure progress and success. Much of what we did involved curating thematic conversations to bring together different perspectives and different approaches.

The main theme of this chapter of my career was learning about the strength of thought and action that can occur when you bring together different actors, especially when addressing a challenging problem. You can start to untangle the challenge and uncover new and better solutions if different people dedicate their best resources and best perspectives.

Pagitsas: What happened when CGI closed its doors in 2016?

Palanza: It was a poignant moment at our last big CGI Annual Meeting, as we gathered for a final time to mark twelve years of Commitments to Action. As we bid farewell to the CGI community, our hope was that our members would take the lessons we had all learned about the strength of partnership, the benefits of cross-sector cooperation, and the need for clear and transparent measurement and disclosure, and continue to utilize and promote those actions in their own work.

At about this time, I received a call from the Bill & Melinda Gates Foundation, and they said, "We're thinking about building a platform to advance progress on the UN Sustainable Development Goals. What should we be thinking about to build something like this?" So, I had the wonderful opportunity to take some of what we had learned at CGI and apply it in a new way for a different community.

My time at CGI and with Goalkeepers, the Gates Foundation program, magnified the importance of the role of the private sector in advancing progress. That's another critical theme, not just in terms of philanthropy— although that's obviously important, especially in the American context—but critically, utilizing a company's core business capability to orient toward or around social and environmental goals. It's the only way we're going to have the scale and the leverage needed to make big change happen globally.

You think about all the levers we have to pull to advance positive impact. Sometimes we need to pull lots of little levers, and sometimes the change needed requires the adjustment of a few big levers. Finance is a big lever. That, frankly, is what kept niggling at the back of my brain every time we curated a big partnership or solution. The question wasn't only about whether the respective partners had enough resources to do what they hoped, but also what role finance was playing. Are the capital markets advancing or hindering this work?

Pagitsas: These questions seem to be a natural bridge to your role at Barclays, where you have the opportunity to address the intersection of the private sector, capital markets, and sustainability in finance.

Palanza: Yes! When the unique opportunity came along to join Barclays, it was candidly not because I'd always dreamed of going to work for a 300-year-old bank. [laughs] In some ways, it was an unlikely or unexpected next step in my career. From a personal perspective, I thought, gosh, this is the chance to better understand and tackle the things that we need to unlock and enable.

I have a skill set unique from that of many people in this bank. Initially, I was intimidated by that, but I thought that what I did know could be of service. Frankly, my perspective as an outsider has been helpful as we've gone on this journey to define, first and foremost, climate change, but then broadly, ESG for the firm. You have to have a big imagination and be able to see the entire universe around you to be able to understand the role that you play within an institution, and that an institution can play in this environmental and social space globally. That's how I've ended up here, and it's been quite a journey.

Pagitsas: Before we dive into defining climate change and ESG for Barclays, would you ground us in Barclays' core products and services?

Palanza: Barclays is a universal bank with a diversified and connected portfolio of businesses, with a 330-plus year history. We support consumers and small businesses through our retail banking services and larger businesses and institutions through our corporate and investment banking services. We also have a cards and payments franchise and a wealth franchise.

Pagitsas: Within this scope of products and services, I am interested in how you define climate change strategy at Barclays. Are there key pillars serving as the foundation for your strategy? Are time-bound quantified goals linked to them?

Palanza: To start, I think it's important to note that our ESG policies and strategy are rooted in our corporate purpose, which talks about deploying "finance responsibly to support people and businesses" and "championing innovation and sustainability for the common good and the long term." Our climate strategy is set with the premise that Barclays, at its core, exists to champion long-term sustainability, so we can and must make a real contribution to tackling climate change and help accelerate the transition to a low-carbon economy. In 2020, we articulated our approach, setting an ambition to be a net-zero bank by 2050—one of the first major banks to do so—and we made a commitment to align our entire financing portfolio to the goals and timeline of the Paris Agreement.

Those are the big picture, time-bound goals. However, it's critical that we demonstrate progress and a sincere commitment to achieving them in the meantime. To put these big goals into practice, Barclays created a methodology called BlueTrack, which sets near and medium-term sector-specific targets against Paris Agreement-aligned benchmarks. We started with our energy and power portfolios because they are the most carbon-intensive and most material for Barclays. By 2025, we will reduce the emissions intensity of our power portfolio by thirty percent and reduce the absolute emissions of our energy portfolio by fifteen percent. We'll be announcing targets for the cement and metals sectors in 2022 and will continue to make our way systematically through our entire financing portfolio, setting science-based targets to work toward our ultimate ambition of net zero by 2050.

Pagitsas: The strategy is quite advanced. How would you contrast what happening today versus when you first started at Barclays?

Palanza: When I first arrived three years ago, ESG hadn't permeated the fabric of the bank yet. Now, I walk into a room, and it's hard to find anybody who doesn't know what ESG is or know at least a bit about Barclays' journey on climate. Of course, some folks will be much better versed in the details around our positions on climate change, deforestation, human rights, etc., but that spectrum of knowledge is to be expected.

When I first got to Barclays, the small team focused on sustainability and ESG worked primarily on ESG reporting and reputation risk assessment. They started the initial steps of building out the bank's first public position on climate change. It was a conversation primarily between the sustainability team and the natural resources banking team. Yet it was not yet a conversation from a bigger, enterprise-wide perspective. The team had the buy-in of the executive team to move this position forward because otherwise, it would not have had sticking power. However, in those early days, it was very much about the players who were working in their silos and coming at it from sometimes opposing views. There were some difficult conversations and some good ones.

I think one of the biggest obstacles that perhaps any sustainability team faces is the misconception that we are here just to stop business. If done well and done correctly, embracing and enhancing the integrity of your ESG credentials as a firm should open up new avenues for commercial success. Sometimes it might take a bit of reorientation away from some lines of business and toward others. That kind of strategic shift is natural and inevitable in a business that hopes to stand the test of time. And in this case, it's good for meeting the existential threat that is climate change and imagining the future of the global economy and the many opportunities therein.

Since then, we have evolved in an extraordinary way to have a bank-wide, embedded approach, which has become much more sophisticated. We have committed to aligning our *entire* financing portfolio to the Paris Agreementv and have set a net-zero ambition on top of that. We have this huge opportunity when it comes to green finance, and we need to make sure that we're building integrity into our climate response as we orient our banking teams to serve that incredibly promising and growing market.

Pagitsas: What was the tipping point internally or drivers to this becoming an enterprise-wide perspective?

Palanza: There were several drivers. First, this kind of change can only happen if the firm's leadership actively champions it. We had that leadership in our board of directors and our executive committee. This doesn't mean everyone needs to hold hands and agree all the time. It is good if there is robust debate and an opportunity to kick the tires, so to speak, to ensure you are making decisions that will work within the institution.

Second, the universe around us has changed in an extraordinary way. This gets into the bigger picture—around the role of the private sector in society. In a relatively short period of time, we've seen a mass societal declaration of expectations of business, shifting to demand business accountability for the wellbeing of a broad set of stakeholders—communities, employees, civil society, and indeed nature itself—rather than just increasing shareholder value.

As we are all more interconnected than ever before, it means that most people have better information and can make clearer demands on the power structures in place—whether that's governments or businesses. And as we are blessed with a lot more access to information, it means that businesses must then make sure that their ESG and sustainability credentials are accessible, transparent, and backed up by facts and data. Happily, I think we as consumers and community members are all a lot savvier about greenwashing, which means corporate sustainability can't just be about vague marketing schemes.

I think the third big driver is that we're living through one of the largest wealth transfers in human history, from older generations to the current ones. Fundamentally these younger generations have core values and expectations of business that are different from what generations before us have had. Part of that is because we now understand the ramifications—and the negative consequences—of some activities that business has undertaken in the world. If you cut down a giant, pristine rainforest to farm cattle, you might be able to sell hamburgers at a profitable rate, but there's going to be consequences, and the more we know what those consequences are, we can hold people responsible for them. And these new investors are certainly going to be making very conscious choices about where their money goes.

Pagitsas: You mentioned that your time at CGI and the Gates Foundation raised your awareness of the importance of financial markets in solving environmental and social issues. What do you view as the role of markets in addressing these issues?

Palanza: Well, as I mentioned earlier in our conversation, I think of finance as a big lever for change. I don't think finance can solve these challenges on its own because it takes the right ecosystem of actors to make lasting, systemic change. But it's an incredibly important part of the puzzle.

Money itself is agnostic. That is, it's just a tool to which we have assigned value, so the power lies in how we deploy it. As a bank, if we facilitate the flow of capital toward innovative climate solutions, equitable housing models, and green infrastructure development, those things can flourish and grow. Yet, if we continue to lean on old financial models—the ones that would tell you that the goal is profit maximization, even at the expense of communities and livelihoods and the natural world—we will never make the change we need to see. We'll never reimagine our economy in a way that allows for *sustainable* growth. I think markets can be an incredible driver for change, but they have to be harnessed and stewarded with a principled perspective at the helm.

We need policymakers and regulators to create the right enabling environment for capital to flow to solution sets. We need the real economy—our clients— to adapt their own business models to transition to a low- or zero-carbon future.

Pagitsas: For Barclays, embracing the transition to a low-carbon economy is key to your ESG strategy. What does that mean, and why is it important in tackling climate change?

Palanza: From the outside, it's easy to articulate a dynamic that there are "the good guys" and "the bad guys." I don't mean this in any glib kind of way. Obviously, climate science has helped us understand that many of the global industry and commerce models we have depended on are not serving us, as they have exacerbated climate change. We all know that we need to move away from fossil fuels to sustainable sources of energy. We know that large-scale monoculture agriculture or animal agriculture contributes to ecological devastation. We know that our transportation systems are outdated, inefficient, and hugely dependent on unsustainable power and energy sources.

When you think about how we are going to make the necessary change we need to see globally, it can't really come from a wholesale on/off switch of our economy. We have to transition the business models of today to reimagine them for tomorrow. In many cases, we need the scale, reach, and experience of the companies which have existed for decades, but we need them to transition all that capability to run their business in a different way, in a sustainable way. And we, as a bank, can support these clients with expertise and best practice—especially with the financing—needed to make the shifts in their own businesses, so they are taking sustainable energy or agriculture or transportation models to scale, rather than their former approaches. Not everyone will be able to make the transition, but many industries can, and we need to support that shift.

Meanwhile, there's a whole future of amazing nascent technologies, industries, and companies that are growing quickly to respond to social and environmental challenges with entirely new solutions. Hopefully, the world has learned its lesson about how long it took for renewables to become competitive in the market. I hope that won't be repeated when it comes to things like green hydrogen or sustainable transportation or other technologies that need to reach scale quickly. We need to grow those businesses of the future while moving the current economy into a new sustainable paradigm.

Pagitsas: What is the primary challenge facing the transition?

Palanza: It means changing a whole system—from the sources of energy currently being extracted and used to produce power, to the transmission grid transporting electricity, to the businesses relying on it to manufacture or produce, and to the individual consumers on the demand side. Right now, in the United Kingdom, people are sitting in extraordinarily long queues for petrol. You realize how dependent everybody is on this system working in the way that they've always been accustomed to having it work.

When you're looking to transition a large and complex system like a global banking portfolio, it's like steering an aircraft carrier. Every degree of change that you can make is a significant shift in the long term, but it's awfully hard sometimes to feel like anything's happening in the near term. It's also both important and difficult to prove progress. You have to see how it's shifting in order for stakeholders and others to believe you that something real is changing.

Pagitsas: Barclays has a relatively large financing portfolio in traditional fossil fuels. How are you evaluating the portfolio in light of the imperative to transition to more sustainable fuel sources?

Palanza: It's a good question. Yes, we have a responsibility and opportunity and a role to play in this transition because we have a legacy financing portfolio that includes a large energy and power portfolio. When Barclays bought the Lehman franchise, it was heavy on the fossil fuel industry. We now have a strategic opportunity to do something about it.

Frankly, the easy answer—or easy on a certain level—would be to get those companies out of our financing portfolio as quickly as possible. Then we look like we have clean hands. So, some people might ask, "Why would you not just cease financing these companies immediately?" The most important factor to consider is what will actually make the change we need to see to halt the climate crisis.

The answer is twofold. First, going back to the concept of transition, we want to be stewards of the transition to support our clients in shifting their businesses to low-carbon and sustainable models wherever possible—for all the reasons I mentioned before. Of course, we recognize that there might be some companies, entities, or even whole subsectors that cannot transition due to the nature of their business models, and we might have to make some different, difficult decisions about them. However, our climate strategy rests on this idea of being active agents of change in the real economy.

Second, if we were to cease financing these energy companies overnight, there is more than enough available capital in the world for them to simply pick up new financing. But will that new financier uphold firmer standards and identify more and better ways to hold those companies accountable? Maybe, maybe not. Therefore, Barclays made the strategic decision to lean in and steward the transition of our entire portfolio to the extent that we can. We have set carbon limits for our energy and power portfolios through the targets we set. Utilizing the BlueTrack methodology I mentioned, we make active decisions about which clients and transactions we can support to meet those targets.

Lastly, and importantly, we have specific restrictions in place about the types of financing we won't do, as a rule, which support and hasten our overall approach. For example, we have rules in place about not financing exploration and production in the Arctic or new thermal coal projects, among many others.

Pagitsas: How are you helping your clients grapple with the transition?

Palanza: We think the highest and best use of a bank of our size is to lean in with our clients. Being a universal bank, we have a wide client base from giant multinational conglomerates to high-growth corporates to local small and medium-sized enterprises [SMEs] to retail customers.

Our business activities range from lending and business banking to investment banking and capital markets activity which gets up into the billions without even blinking. Our mortgage portfolio will be an important area of focus for us as well because residential real estate is a critical sector to decarbonize, particularly as the UK government has set it as an area of focus for its own net-zero strategy. We will be setting carbon reduction targets just as we will for other high carbon sectors, as part of our commitment to align our entire portfolio to the Paris Agreement.

Each of these sectors and client types has its own challenges to grapple with. I know fossil fuel companies make the headlines, and there's no doubt dealing with the energy and power sectors is key to facing the climate crisis. But some of the sectors hard to abate—that is, those for which there is no easy technological solution to transition, such as cement, steel, or chemicals companies—are facing huge questions.

Meanwhile, SMEs of many sorts are grappling with a lack of resources to measure and disclose their ESG credentials. At the same time, all companies see a fragmentation of reporting frameworks and expectations across markets and geographies, so it's difficult for them to share what they are doing in a way that ratings agencies and investors can analyze and compare.

Pagitsas: If your clients aren't able to make the transition, which may not be green enough or fast enough for some people, what does that mean for the climate change issues that we're talking about? If you abandon those clients who can't or won't transition, what will happen?

Palanza: If the real economy—including our clients as well as everyone else's—can't or won't transition fast enough, we won't be able to decarbonize our economy fast enough. We won't be able to stay below the warming threshold that climate scientists have told us we must adhere to or face devastating planetary changes. We have already far exceeded the earth's carbon budget, so we need to—collectively, all of us—do everything we can to cease additional carbon emissions and actually work on carbon removal from the earth's atmosphere. So, banks have a critical role to play in an entire system that needs to shift, and quickly. The concept of the transition might feel weak or gradual, but we are talking about a complete reimagining of our economic system, and it does have to happen at the rate which climate science dictates.

At this point in time, if we walk away from clients who can't or won't transition, as we discussed earlier, I think most likely they will go to another bank or financier. But the entire finance industry is changing quickly. There's no doubt there is still private equity and debt; other private financing that's stepping in where some of the banks are exiting. Because of the collective pressure of our stakeholders, I believe that the entire finance sector is going to have to move.

Eventually, I think the whole finance sector will need to set carbon reduction targets backed up by data and transparent disclosures, where they can be held accountable. And when you account for more suppliers of private capital, it starts to matter a lot. As I mentioned, in the near term, I think banking clients could potentially go elsewhere and find the capital they need to do things which make climate change worse, things like building new thermal coal-fired power plants. However, even if you looked at this from a purely economic perspective, it will come down to whether those assets have the potential to become stranded and a financial liability down the road.

Pagitsas: A "stranded asset" is a financial investment that has become a liability, usually due to premature write-downs or devaluations. Assets that are vulnerable to climate change, such as a power plant located on a hurricane-impacted and flooding-prone coast, might become a stranded asset. What does "down the road" mean?

Palanza: Yes. Climate used to be considered a purely reputational issue. We know it is a threat to life on earth and is a massive, global societal challenge. But policymakers and researchers—including Mark Carney when he was the governor of the Bank of England—rightly identified climate change as a financial risk. So, there are sectors and industries which, if deemed to be out of step with the transition, might suddenly see the bottom drop out. Or, in the example you gave, the changing environment itself might pose physical risk, becoming a threat to certain assets.

"Down the road" used to mean three decades from now. Now, "down the road" could be right around the corner. The time factor has changed dramatically for all of us. That is a big part of what has made people wake up and realize that this risk calculation is not something that can be dealt with under another administration, another CEO, or another whomever. It has to be calculated and acted upon *now*.

Pagitsas: Driving solutions *now* was a big focus of the November 2021 UN Framework Convention on Climate Change [UNFCCC] Conference of the Parties [COP] in Glasgow. As a planet and a society, we no longer have the luxury of time to address climate change. What key conversations or decisions on climate change did you and other private sector partners drive toward at COP26?

Palanza: As you identified, COP is designed to be a government conference. While the private sector has been interested in the policies being negotiated at past COPs, I don't think business has ever shown up in force the way it did this year—the finance sector in particular.

We were delighted to be founding members of the Net-Zero Banking Alliance [NZBA], part of the Glasgow Financial Alliance for Net Zero [GFANZ]. It is an important and far-reaching alliance of financial actors who have all set net-zero goals and are committed to working together, not only as a coalition at COP, but going forward. We were co-leads of the Financial Services Task Force Net Zero Working Group, part of the Sustainable Markets Initiative launched by His Royal Highness The Prince of Wales, which produced a practitioner's guide to share practical experience with banks looking to set their own net-zero goals. We also participated in working groups focused on carbon markets and committed to the Get Nature Positive initiative, among others.

I could give you a complete laundry list of partnerships and coalitions we signed on to, but I think the important point is the fact that there's no going back. The finance sector—and the private sector, broadly—is taking big steps forward, with the whole world bearing witness to its progress.

Pagitsas: What happens now after Glasgow?

Palanza: We have set our big commitments. So, we are committed to following through on those! Events like COP are helpful deadlines and crescendo moments, which the global community can use to push for agreement on critical policy issues. However, you're right to ask about after Glasgow because it's important to acknowledge how much work there is still left to do. We have more sectors to align to BlueTrack, targets to set, and details to work out about how best to gather the right data and the right methodological approaches to calculate all our financed emissions.

We have committed to aligning our *entire* portfolio so that work will continue in earnest. But we are also keen to continue working with our peers to advance the banking industry's ability to support clients and work together to move the whole industry forward on climate, particularly in our ability to measure and assess our financed emissions.

Pagitsas: Why is partnering with peers an important part of Barclays strategy on the climate change agenda?

Palanza: I feel fortunate that Barclays is as big as it is because we have the resources and buy-in to be able to make these changes internally. I sit on the Principles for Responsible Banking [PRB] board of the UN Environment Program Finance Initiative [UNEP FI]. It gives me a good opportunity to learn from my colleagues who work for other global banks as well as smaller, regional banks in Latin America, Scandinavia, and North Africa, for example, where they have different sets of issues at play—different challenge but also different strengths, including different stakeholder expectations.

That's where these consortiums of financial players are important because we can take the lessons that we're all gathering in real time in different contexts, and everyone can learn. We debate questions such as, "How would you set a net-zero strategy that has integrity? What factors do you have to take into account? How do you start measuring financed emissions in every part of the economy, when none of us have perfect tools?" Those kinds of questions are the things that matter when it comes down to the brass tacks of setting your own strategy as a bank, what that means for clients, and what it all means for our chances of making real impact on the climate crisis.

Pagitsas: As interest increases in ESG and sustainability and engaging through coalitions, how would you advise companies entering the ESG and sustainability space? How should companies balance cooperation and competition with industry peers in the same coalition?

Palanza: If a company wants to think about how to take their first steps in developing a sustainability strategy, I'd say start by looking at your core business. What are the issues that are most material to you as a company, and what are the issues driven by your business, which are most impactful on the world around you? A credible ESG strategy should hang off of your core business strategy. It should not be a tangential or separate approach.

Getting sector-level intelligence from peers is an important part of the equation for any company seeking to get up the ESG learning curve. If you can look to your left and look to your right and understand what your peers—and indeed competitors—are doing, you can understand where you need to go in order to have a differentiated strategy.

Public disclosures have become better and stronger. There's a long way to go there still, but it's helpful to start reading with a discerning eye what other ESG reports are saying and determine what they actually mean. Are they setting very specific KPIs [key performance indicators] and targets? Are they backing those up with actual evidence, and describing outcomes and impact? Are they showcasing progress in a discernible way and not behind a veil of fancy marketing language? It's a wonderful thing if more and more of our competition is based on who has the strongest ESG credentials.

Explore whether there are consortiums to join at an industry level to work with your peers to advance your knowledge and capabilities. There will always, of course, be a competitive aspect to this, but the opportunities to learn and share best practices that come from these industry coalitions are extraordinarily helpful. There are no definable boundaries to determine whose climate action impacts whom, as the climate is a planetary system. So, that means it will quite literally take all of us working in coalition, or at least in our own little corners, to have any hope of getting where we need to go. In the case of Barclays, it's helpful to get together with our cadre of banking peers, especially those who have big capital markets activity like us, to talk about the role of our capital markets activities in financed emissions. It's also helpful to talk to a variety of other banks and financial players who are dealing with very different issues because we play a collective role in addressing climate change and in transition.

Lastly, you can't forget the really important role of cross-sector learning and collaboration. We have gained a huge amount of knowledge from conversations with NGO experts, academic institutions and scholars, and government leaders.

Pagitsas: What is going to be the biggest change for the banking sector over the next five years? What do you hope for more broadly in society?

Palanza: We're seeing a huge rise in fintech innovation, and even general technology companies dabbling in banking. This is an interesting development because banks as institutions have engrained societal contracts with the communities in which they work. Fintech companies don't necessarily have the same connection to place and community, but they offer loads of accessibility and ease in transacting. So, I think the real question is who will succeed in securing stakeholder buy-in and trust. That's a big question for the banking industry.

At the same time, you see a race to the top by banks and the private sector at large in terms of articulating more and better plans to meet society's expectations. We've seen COVID-19 unearth and reveal appalling societal challenges that have been there all along but that perhaps we just weren't required to face at a larger community level.

What's the role of the private sector in promoting equity and social justice? Society has, rightfully, less and less patience for a company's answer to that question being encapsulated in a one-day celebration or in buying the rights to a particular cause-related march. Business needs to integrate those principles into the way it treats its own employees, stand up for them in the communities in which it works, and—in the case of a bank—be true to them in its financing choices.

We've talked a lot about an aspect of the *E* in ESG because climate has dominated so much of the conversation inside banks. But we can't be siloed in our thinking. We live in a complex, interconnected world, and the challenges we face are matrixed and messy. You can't think about climate without thinking about the human systems we have developed, the communities most impoverished, disenfranchised, or most impacted by climate change, and the livelihoods dependent on a shifting economy—the *S* in ESG. We can't think about climate without thinking about nature and biodiversity at the core of our planetary system—the natural world on which we are all quite literally dependent.

Hopefully, if we can evolve to account for our holistic impact—both positive and negative—and if we can learn from one another's expertise and perspective and work together across sectors and disciplines, we might have a better shot at solving the biggest sustainability challenges facing us today and be better prepared for those facing us tomorrow.

Yulanda Chung

Managing Director, Group Head of Sustainability, Institutional Banking Group
DBS Bank

Yulanda Chung is the head of sustainability at DBS' Institutional Banking Group. DBS is a Singaporean multinational banking and financial services company with operations in 17 markets and 650 billion Singapore dollars in assets at year-end 2020. She leads the bank's responsible financing agenda and advises on environmental and social issues pertaining to transactions.

Yulanda was previously head of sustainable finance at Standard Chartered in London, where she established the bank's sustainable lending agenda. Before joining the banking sector, she was an equity analyst for the mining and building materials sectors at RobecoSAM in Zurich, Switzerland.

Yulanda is a trustee at the think tank TransitionZero, which provides financial analytics in the power and heavy industries to support decarbonization and is a founding partner of the Al Gore-backed coalition Climate TRACE. She is a former board member for the Association for Sustainable & Responsible Investment in Asia (ASrIA), and during her tenure, the organization was integrated into the United Nations Principles for Responsible Investment.

Chrissa Pagitsas: Today, you lead sustainable finance for DBS, a major financial institution. However, you began your career in journalism. What was your professional journey to your current role?

© Chrissa Pagitsas 2022
C. Pagitsas, *Chief Sustainability Officers At Work*,
https://doi.org/10.1007/978-1-4842-7866-6_19

Yulanda Chung: I started my career as a journalist, which was a childhood aspiration. I fulfilled it when I started as a cadet for the *South China Morning Post* in Hong Kong, which is a widely known newspaper in this part of the world. As a cadet, I moved from desk to desk covering everything from local news to business news, China news, features, and photography. After the cadetship, I chose environment and consumer affairs as my beat. That set the foundation for my career in sustainability. I researched and interviewed people about consumer affairs and the environment and how the two often have convergent interests and conflicting challenges.

In 2000, I moved on to become the correspondent for a magazine called *Asiaweek*. At that time, *sustainability* and *finance* were two words that did not come together in any shape or form. It was unheard of. I had the good fortune of interviewing a woman who is considered a pioneer in sustainable investment, Tessa Tennant. She became my very dear mentor, and she co-founded the CDP to promote sustainable investment.

Because it was twenty years ago, no one had heard of sustainable investment. It was very difficult to translate "sustainable development" into Chinese because it's quite a chunky phrase when you say it in Chinese as a term. During my interview with her, I was awestruck by the notion that sustainable development can be promoted by working in finance and why finance plays such an important role in driving sustainability.

I quit journalism, really, for her and the course that she was charting as a pioneer. After I worked for her at ASrIA, she encouraged me to pursue my career in sustainable finance. She introduced me to RobecoSAM, a sustainable asset management company, which was another pioneering firm at the time and based in Zurich. It pioneered the development of the Dow Jones Sustainability Index [DJSI]. I worked for them as an equity analyst and learned how to create valuation models. This set the foundation to integrate sustainability analysis into financial analysis.

This was the early 2000s when the DJSI was still in its early days. Yet already I could see that companies were vying to be included in the index. Some of the companies, the ones with the foresight to know that this was going to become big, were already making a lot of inroads to make sure that this index formed the basis of a lot of investment funds with a sustainability focus. They expected the assets under management [AUM] of these funds to grow and grow and grow. I could see how finance had the potential to promote better sustainable behavior at corporates.

I oversaw the mining and metals and building construction materials sectors. One of my favorite companies that I covered was the water sanitation company called Geberit, a Swiss company. They invented the dual-flush mode for toilets, with the option to choose the small flush or the big flush. That was novel at the time.

Because my first degree was in political science, I wanted to deepen my training in the environmental area. I went back to school and got a master's degree in environmental change and management. With that academic training, I went to work for an environmental consultancy in Hong Kong, ERM. I learned a lot there doing a lot of technical work. I also did a lot of manual but essential work such as collecting soil samples and counting trees and dolphins.

After a while, I wanted to go back to what I did before, which is marrying finance with sustainability. Fortuitously, the opportunity came up with Standard Chartered, where I was hired as their first sustainable lending manager. This was in 2007 and the early days of sustainable finance. At that time, a lot of the effort went into convincing the bank that there was a need for such a role. However, Standard Chartered was one of the early signatories of the Equator Principles.

It is important to acknowledge the Equator Principles initiated the focus on environmental and social [E&S] due diligence for project finance transactions. The agenda thrived because it was narrowly defined to investigate the impact of a physical asset with a relatively well-defined boundary. Once you've cracked that and built a foundation, you can branch out. Nowadays, the environmental and social risk management regime has become more sophisticated, covering climate risk, physical risk, transition risk, social risk, and governance roles and responsibilities. E&S risk management is not confined to project-related financing either.

At the end of my time at Standard Chartered, I re-located from London to Singapore. In Singapore, I worked for a think tank called IEEFA, the Institute for Energy Economics and Financial Analysis. I had a great time doing research for them. Understanding the power sector in Indonesia was my area of specialization. I moved on from there to where I am now at DBS. I think my career history is a bit eclectic—journalism and time at a non-governmental organization, asset management and then consulting, and finally commercial banking. However, the common thread that runs through my journey is the environment and using finance to promote sustainable business practices.

Pagitsas: Another common thread seems to be the creation of new structures and launching new businesses related to sustainable finance.

Chung: Yes, that's a very astute observation. When I was hired at DBS, it was a relatively blank canvas. The DBS leadership hired me because they realized that times have changed, and the writing was on the wall to make structural changes to how sustainability was viewed through a financing lens. Regulatory and investor pressures were also increasing. I have been very fortunate to have the opportunity to build the sustainable finance agenda and to set-up teams with the necessary skillsets. In many organisations I've been in, I would usually be hired as the sole person working on the agenda and then bring more people into the team as the agenda became more entrenched in the company's business model. As a result, training people on sustainable finance has become a big part of my career.

Pagitsas: Given that you have been working in sustainable finance for more than twenty years, what are the biggest changes you have seen in the market during that time?

Chung: Back in the early 2000s, I saw how important it was to many companies to be included in the DJSI and how that inclusion drove strategies and then change. At the same time, it was hard going as we faced a lot of hesitation to change business-as-usual practices. When I worked for ASrIA in Hong Kong, we spoke to many local asset managers. Even though many of them showed interest, that was really it. Nothing substantive happened. Fast-forward to 2020, 2021, and the only investment funds that have seen significant inflows are ESG [environmental, social, and governance] funds. I feel that after twenty years, maybe, the time for ESG investment has come.

Today, I am surprised by how much interest and action is dedicated to the ESG agenda. Over the course of twenty years, I believe that the role of finance in sustainability issues has increased in importance. I continued on this career path in sustainable finance, not thinking that it would become big. It was more that there were new things to do all the time, and as a result, I kept going.

Just as with my own career trajectory, the cause itself has moved. It started in Europe, where I had my first professional experiences in Zurich and London. Now it has come to Asia, which has caught up in no time.

Pagitsas: Indeed, your professional experience has taken you from Asia to Europe and back to Asia. Do financial institutions or companies define sustainability and ESG differently in the Asian markets versus the European and North American markets?

Chung: When I first talk to clients and bankers about sustainable development or sustainable finance, the discussion often goes through similar stages regardless of where the clients are located geographically. First, they find it rather novel and very interesting. They ask a lot of questions, which are then followed by statements such as, "Oh, it's never going to work because there are data challenges. And, with imperfect data, you really should not do all these poor analyses that you're doing." Then it goes on to the next stage, whereby people realize, "No, I think I'm the outlier here." This change happens because they realize sustainability is no longer just icing on the cake. It is simply part of a strategy to future-proof your business. Whether I'm speaking to clients in Europe or in Asia, they all go through these stages of understanding.

However, the difference is that the pace of adoption is different. What took me seven to eight years at a European bank to build with sustainable finance,

I have done in three years at DBS. Nowadays, I advise customers on the dual materiality issues that resonate with them. This is common for customers across markets. Dual materiality is an extension of the accounting concept of materiality of financial information. Information deemed material should be disclosed. The "duality" comes in because not only can ESG-related impact be material on the company's financial ratios, but also the impact of a company's operations and core products or services can be material on the environment and the society.

Pagitsas: I'm hearing several things in parallel. First, regardless of the geography, the education phases are the same. Second, the pace at which we're going through the educational phases is much faster today than it was ten years ago, twenty years ago. Third, the issues customers in different geographies face may be different in nature, but the principle of dual materiality applies equally.

Chung: Exactly.

Pagitsas: An important tool that financial institutions have today is transition finance. In July 2020, DBS launched a sustainable and transition finance framework and taxonomy. What is transition finance and the scope of the framework and taxonomy?

Chung: At DBS, we define transition finance as financing that supports decarbonization activities by our customers in line with the time-bound nature of mitigating climate change and takes into account the total environmental, social, and economic impact of the activity throughout its lifespan. The nature of the transition in each country or region is influenced by the evolution of the entire system, including local strategies and policies.

Many companies in the developing world are only making incremental changes because the alternatives are not commercially available yet. They can't jump into hydrogen and large-scale battery storage which will allow the full transition to a low-carbon economy. There's no point in trying to deny them the pool of sustainable finance that is growing. You want to mobilize that pool of sustainable capital to enable incremental but instrumental change.

Transition finance needs to be understood better. It's about transparency. That's our motivation when we talk about transition finance. This is because there is controversy around it. Questions arise. Is transition finance a copout? Is it inferior to green finance? Why are you doing this as a transition deal instead of financing dark green measures? It is not that black and white. It's a spectrum. The transparency comes with our client saying, "I may own a gas company. I'm a refiner, and I am looking at reducing my methane emissions. However, I'm still transporting natural gas." Nevertheless, we need to be clear that exploration and development of new fossil fuel projects are not transitioning when you look at life cycle impact.

The next natural question about transition finance is, "What is the destination?" When people talk about transition, they have to think about if the end goal is net zero. If it is feasible, by when? Is it 2050? Or maybe in China, it could be 2060. If it is going to be 2050 in Europe, can we realistically demand having the same deadline for emerging markets?

Pagitsas: There may be two timelines for us to clarify. You're pointing out that people globally are moving much more quickly through the educational phases of sustainability and sustainable finance. However, emerging markets such as those in Asia may not be able to execute and deliver on sustainable goals and activities according to the same timeline as the European markets.

Why can't emerging markets move any faster than 2050 or 2060?

Chung: I think that's right. To answer your question, the pace of transition differs in different regions partly because when we talk about promoting sustainable development, it's not just about finance. It's not just about the private sector. It's also about the regulatory regime. When you don't have a regulatory framework enabling those changes, it's a lot harder. As a multinational corporation, what you do in North America compared to what you do in Indonesia will be subject to different regulatory requirements and enforcement. That plays a part.

Pagitsas: What is an example of a transition finance deal that you have worked on for a multinational client in an emerging market?

Chung: I have a couple of examples. One transition finance deal we did was a sustainability-linked loan structure. This structure allowed companies to borrow money and use it for general working capital. However, it is not like a green loan where the entirety of the loan had to go to the earmarked eligible green projects. That's one difference between a green loan and a sustainability-linked loan.

At DBS, we also work with a marine infrastructure company called Sembcorp Marine. They provide engineering solutions and build the infrastructure for the global offshore, marine and energy industries. They realized early that the oil and gas industry was undergoing structural transformation with energy transition arising from climate change considerations, and there was a need to diversify into new sustainable business solutions. Sembcorp Marine's group finance director, William Goh, has placed great emphasis on proactively evaluating its strategy to develop new revenue streams. Two examples are the design and construction of offshore sub-station topsides for offshore wind farms, and zero emission battery-operated passenger and vehicles ferries, where green hydropower is used to charge the batteries. Presently most ferries use marine gas oil as bunker fuel that is more carbon intensive. They are also developing LNG-battery [liquified natural gas-battery] hybrid tugs to promote sustainable port and yard operations and pioneering the use of floating foundation solutions for next-generation offshore wind turbines.

Concurrently, the company is also increasingly adopting greener approaches to execute its orders. These include installing solar panels in their yards and using the renewable energy to power their equipment and renewing its forklift fleet from diesel-powered machines to electric-powered ones to reduce the overall carbon footprint of its operations.

You can see that my counterparts with whom I discuss sustainable matters are not confined to environmental, health and safety professionals, but include finance leaders such as Goh who see sustainability as a strategic driver for performance.

With all these activities, Sembcorp Marine is still servicing the fossil fuels industry. One may ask, "Is this green enough?" I think that it's a good illustration of transition finance. We're being very clear about what we are doing and what those improvements are. In sum, the company is indicating that with all these activities, it has a whole package of changes. Some of them are zero-carbon in nature and some are not yet, but the importance lies in the company's clear understanding of the steps necessary to tackle the clean energy transition head-on.

For them, we were able to set the performance targets for the company. If the proportion of their overall revenue derived from these earmarked sustainable business revenue streams meet the target defined upfront, then they will be able to earn a discount in the interest margin from the loan that we extend to them. That's how we try to encourage and incentivize bolder steps from our clients. It incentivizes the client to think, "Okay, my cost of financing is tied to committing to responsible business solutions, and by doing more we can achieve a win-win outcome for stakeholders."

Pagitsas: Turning to DBS, what is the scope and pillars of DBS' sustainability strategy?

Chung: The first pillar is responsible finance which is our core pillar. In this pillar, we integrate environmental, social, and governance considerations into our financing decisions. We allocate capital toward sustainable economic activities while staying vigilant to manage the environmental and social impacts arising from such development. This is also where the bank addresses financial inclusion because, in Southeast Asia, which is our core region, there are many vulnerable communities, such as migrant labor.

In Singapore, for example, some migrant workers may not have access to bank accounts or be able to send remittances home because the transaction costs are likely to be higher in servicing them than the revenue a bank would likely earn from them. DBS made a very conscious effort to serve this market to make sure that the migrant workforce has access to financial services and can remit their hard-earned money home.

The second pillar is our own shop and how we responsibly manage our operations. We have developed circular economy procurement principles and use them to manage our suppliers. Our property management team manages all our branches and premises and recently even refurbished an existing building into a net-zero one.

The third pillar is our social purpose, creating positive social impact, which we support through the DBS Foundation. The foundation provides grants to nurture social enterprises in the markets that we operate in, not just in Singapore but also in India, Indonesia, Mainland China, Hong Kong, and Taiwan. These social enterprises are started by entrepreneurs who have a mission to provide solutions for a particular social or environmental ill. Besides grantmaking, as these social enterprises become bigger, they need a wider range of banking services such as financing from a bank. To address these needs, we have a social enterprise banking package for them. The objective is to grow with them through their various stages of expansion.

In 2020, at the height of the global COVID-19 pandemic, DBS approved over 10,000 collateral-free loans totaling more than 5 billion Singapore dollars to small- and medium-sized enterprises in Singapore, with over ninety percent of the loans going to micro- and small-sized enterprises.

Pagitsas: How do you view the increased focus on transparency around banks' performance on these environmental and social issues?

Chung: I am supportive of increased transparency. Banks, or corporates in general, need to adapt to a changed mindset when it comes to disclosure. Businesses tend to focus on their own performance and comparative progress year over year. Now, businesses need to also ask questions about their performance against the planetary boundaries such as, "Is my company going to contribute its fair share to arrive at net-zero emissions?" For instance, DBS' TCFD [Task Force on Climate-related Financial Disclosures] reporting illustrates how the bank drives transparency and openness around climate change risk management. We issued our first TCFD report in 2018. DBS was the first bank in Singapore to sign up to TCFD. It was difficult to create the report because banks are expected to report their scope 3 emissions. Scope 3 emissions are not our own operational emissions. It's our customers' emissions. If our customers don't report their emissions, we don't have much to report.

An additional challenge with carbon emissions reporting is that there are different ways to model the assumed emissions and differences in opinion in how good they are. However, we decided to take the plunge because even though the data is imperfect, or at times, even unavailable, we had to start somewhere. Perfection can deter change for the better. Sometimes you have to invest in learning and doing to improve.

I think that's why DBS has been championing TCFD from the beginning. We knew that it was going to be a learning journey. We are setting an example because DBS is the flagship bank in Singapore and will hopefully raise the bar for other banks in the region to follow suit.

Pagitsas: Why is TCFD an important reporting tool for DBS?

Chung: The emphasis of TCFD is that the data reported is still financial reporting. It moves us away from thinking that it is "non-financial reporting." Climate-related financial reporting is going to affect the financial standing of companies. There's no point saying, "This is financial reporting," and then, "This is the non-financial reporting." If you have that division, immediately, people will be dismissive of the latter. They will say, "You're telling me it's non-financial. I don't need to worry about it. It doesn't affect my bottom line." The change in emphasis is important.

Pagitsas: Clearly, you view the unknowns of climate reporting to be something to tackle and solve rather than to avoid. How would you advise the head of sustainable finance at another bank who has a different perspective? They may say, "We must have perfect data to calculate our scope 3 carbon emissions," or "We can't assess the impact of climate change until we have three years of complete data."

Chung: I would point to the Equator Principles to demonstrate that a sustainable finance agenda can thrive on its narrowness. With TCFD, even though the data requested for reporting is very broad, you can also draw a narrowly defined boundary around the better data. Therefore, my peer at another bank could start the process by saying, "I'm not going to look at my entire financial portfolio. I'm going to look at individual sectors." We know that data for some carbon-intensive industries, such as power generation, is better. When you start with that approach to data collection and reporting, you can make gradual improvement year after year.

At DBS, we took the same approach when we first conducted the climate risk scenario analysis in TCFD to look at how changes in regulations and physical risks are going to affect the portfolio and the probability of default of our customers. We started with transition risk because it is easier to model with the increase in carbon price being used as a way of modeling the change. In comparison, we found physical risk to be more difficult. This is because the location of the assets affected by climate change is relatively imprecise at this point.

We are still learning at DBS. I would share with my peers who are also on this sustainability journey that that you may be pleasantly surprised by the results, which can inform your strategy and inform your capital allocation. A robust sustainability framework is also a way of risk management.

Pagitsas: The challenges you addressed above are challenges related to data and modeling and about being a leader and overcoming the fear of the unknown in this field. Let's talk more about your leadership philosophy. Is there somebody in your career that influenced you as a senior executive?

Chung: My mentor Tessa Tennant had been a great influence on me. Tessa co-founded the CDP in 2000. She had the vision and saw early on how important data would become. Her work paved the way for Mark Carney and Michael Bloomberg to come out with TCFD in 2015. I still remember helping CDP in the early days by licking the envelopes for the letters going to the CEOs of companies and asking them to report publicly. With COVID-19, sealing an envelope is never going to be the same again!

Another aspect I learned from her is that it is important to be an effective communicator. While deep know-how about a subject matter is important, you must articulate it well. Therefore, even early in my career, I made a lot of effort to work on my communication style. When you want to bring your whole company, country, market, or industry with you on a certain cause, you need to communicate well. I learned that from her. She was a terrific communicator.

Lastly, I would say it is important, although it sounds cliché, to be a good listener. A phrase from a coach has stayed with me over the years. He said, "A lot of people listen to respond. They're not really listening." This resonated with me because I thought I was listening, but in fact, my mind was churning with, "How can I come back with a clever answer? How can I come back so that it shows off my knowledge?" I wasn't really listening. The coach said, "No, you should not listen to respond. You listen. That's all." That phrase did wonders for me.

Pagitsas: How have you partnered with peers and senior leaders to move forward the sustainable finance strategy you have set?

Chung: You need to demonstrate to senior leadership the value proposition of sustainability. I know "value proposition" is a much-used phrase, and many of the value propositions I've heard over the years are unconvincing. For a long time, one of the value proposition arguments to the board was, "Sustainability is important because your staff will become more loyal." People think, "You say that, but I don't really buy it."

The value proposition needs to be convincing. First, you need to make sure that the leadership realizes there will be compliance pressure, not just compliance for now but compliance with anticipated changes and compliance with your export markets, for example. That may be a low bar for some people, but it is not a given. Second, you use sustainability to drive performance and employee engagement. The penny has dropped. Indeed, these days, younger staff members tend to be more loyal to employers whose values they agree with.

However, loyalty is can only be earned through authentic action. Therefore, it's not just about sustainability. Let's put that word aside. It's just about where you stand. Where do you stand on certain causes that are important to me as an individual? It is important to be authentic and think about what the company stands for. You will have to be able to defend it. And even so, you will still be challenged because you will not find agreement everywhere.

No company can please all the stakeholders all the time. As a result of some of the inauthenticity the younger generation has seen, they are placing emphasis on values. It is another way of telling senior management, "You need to think about your positioning on issues that you think are unrelated to your core business." Values based decisions and walking the talk may have a great impact how you're going to be able to attract and retain talent. That is now becoming even more important and more of a reality.

Pagitsas: What do you either hope or predict will be different about sustainable finance in five years?

Chung: I'm afraid that I might end on a rather pessimistic note. [laughs] I think our future is informed by our past performance, unlike investment funds, where past performance doesn't tell you much about the future. With the UN Conference of the Parties [COP] convening, we've learned that international governmental collaboration is very challenging, and we may not be able to crack it in time. Without international governmental collaboration, I fear that developing countries will disproportionately suffer more, both in terms of physical risk and economic growth.

Also, looking at sustainability as an industry or as a profession over twenty years tells me that this is the biggest talk shop ever. [laughter] I have never encountered another industry with so many conferences, workshops, and webinars. People fly around the world to do these sustainability conferences. I am guilty of doing quite a few of those. But the change that you gain from talking at these conferences is often not commensurate with the emissions that go into them, or the time and effort spent preparing for them!

On a more positive note, I will say that we know that human history is not just our generation or even my children's generation. It is for many, many generations to come. I believe in the Gaia hypothesis and theory by James Lovelock. The Gaia theory is that living organisms on the planet interact with their surrounding inorganic environment to form a synergetic and self-regulating system that first created and now perpetuates the climate and biochemical conditions to support life on Earth. I have faith in the planet's ability to find its equilibrium. It is capable of healing itself.

Heidi DuBois

Head of ESG
AEA Investors LP

Heidi DuBois is the head of environmental, social, and governance (ESG) at AEA Investors LP (AEA). Founded in 1968 by the Rockefeller, Mellon, and Harriman family interests and S.G. Warburg & Co., AEA is a global private investment firm with over $14 billion in assets under management and an integrated investment platform that supports leading middle-market companies with private equity, growth capital, and private debt solutions.

Before joining AEA, Heidi was global head of ESG at Edelman. In this role, Heidi worked with boards and senior executives to design, execute, and communicate their ESG strategies with investors and other stakeholders, including employees, regulators, and non-governmental organizations.

Prior to her role at Edelman, Heidi was ESG director for the Society for Corporate Governance, where she developed new programs and managed thought leadership and engagement on emerging ESG trends for boards of directors, business leaders, and corporate governance professionals. Prior to that, she oversaw the corporate social responsibility and philanthropic functions at BNY Mellon. Earlier in her career, Heidi practiced law at Debevoise & Plimpton before serving in positions in governance and ESG at Assurant and PepsiCo.

She serves on the ESG advisory board for FiscalNote and the advisory boards for the Business in Society Institute at Berkeley Law School and the GreenFin conference.

© Chrissa Pagitsas 2022
C. Pagitsas, *Chief Sustainability Officers At Work*,
https://doi.org/10.1007/978-1-4842-7866-6_20

Chrissa Pagitsas: Let's begin with AEA, a private equity firm. What are AEA's lines of business and how does your role as head of ESG intersect with them?

Heidi DuBois: AEA has been investing in middle market businesses since its founding in 1968. Five decades later, our global, integrated investment platform continues to support companies around the globe with private equity, growth capital, and private debt solutions. Our private equity strategy supports middle market, small business, and early-stage companies to accelerate and execute on their growth initiatives, operational change, and strategic transformation. The debt group, which provides middle market debt and mezzanine funds, partners with middle market private equity firms, companies, and lead arrangers of senior secured financings to source and provide debt facilities in support of buyouts, recapitalizations, add-on acquisitions, and portfolio re-financings.

AEA's approach to ESG and responsible investing has always been about integration, from integrating ESG across our portfolio and internal operations. We have also been looking at opportunities that proactively contribute to a more inclusive and sustainable economy. Most notably we recently supported the formation of AEA Bridges Impact Corp, a special purpose acquisition company, with Bridges Fund Management, that will invest through the lens of the UN Sustainable Development Goals [SDGs].

I joined AEA in January of 2022 in a newly created role that recognizes the value of effective ESG management and reflects AEA's mission of building businesses for long-term success. In my role, I am responsible for expanding and deepening integration of our existing ESG platform throughout internal and portfolio company operations to support what we already bring to the table in our investments. Our priorities are to ensure that ESG considerations remain a core component of our due diligence processes, our partnerships with our portfolio companies to ensure operational excellence, and our commitment to deliver superior performance to our investors.

Pagitsas: Prior to joining AEA, you led the ESG practice at Edelman, a global consulting and communications firm. What services did you provide to Edelman's clients?

DuBois: Edelman partners with businesses and organizations to evolve, promote, and protect their brands and reputations. Their principal tool is communications, which they use to help clients explain what they do, clarify their mission and values, and build trust with their stakeholders and the public.

Pagitsas: How can companies build—or lose—trust through their communications?

DuBois: There are a million avenues for communicating today. While one must-do is to engage in formal reporting in accordance with recognized ESG frameworks, investors and other stakeholders obtain ESG information in a variety of ways. Edelman's *2021 Trust Barometer Special Report: Institutional Investors* found that investors use traditional channels such as ESG reports, but also look to earnings calls and social media. That means companies need to know how to communicate through different channels. For example, LinkedIn and Twitter formats are very different from the company website, which in turn is very different from a PDF or printed ESG report, an animated presentation you can watch on a mobile device, and from an organization's mandatory regulatory filings.

The challenge with an ESG communications strategy is that you have to be consistent across the board because that's the only way you're going to be credible and build trust with stakeholders. You also need to be transparent across the board while telling your story in a way that works for the particular channel.

Pagitsas: The takeaway is that companies today need to create a surround sound strategy for their ESG communications, like a 360-degree stage at conferences. It's not enough to put out an annual ESG report. You need to intersect with many stakeholders in many different formats. Is that right?

DuBois: Yes. When I think about communications, it used to be one-way communication and then two-way, and now it's every way and in many directions. For example, investors may look at Glassdoor.com to see what employee sentiment is about the company, and they can incorporate that information into an assessment of its culture.

Pagitsas: What are some themes that you're seeing organizations tackling around social issues?

DuBois: Diversity, equity, and inclusion [DEI] are the top social issues today. I also see a shift in the assumption that S issues are inherently more difficult to measure. Edelman's *Institutional Investors* report also found that fifty-three percent of US investors lack full confidence in the accuracy of the information that companies disclose about their diversity and inclusion goals and pledges, leading all ESG categories where investors expressed concerns. In the 2020 report, a large majority of institutional asset managers globally indicated that they were putting companies that weren't making progress in this area on watch lists.

Clearly, this means that work still needs to be done on DEI measurements. In many respects, measurement of DEI is in its early stages because, while some companies have collected self-identified data for some time, many are implementing the necessary processes and procedures for the first time. It may be some years before a fully fleshed-out, measurable social platform has been created.

Pagitsas: If the collection of most social data is in its infancy, does that also mean that companies are still in the early days of understanding the outcomes of DEI policies such as increased diverse representation in C-suite roles and feelings of inclusion for employees of all backgrounds?

DuBois: Yes. About a year ago, I heard a board director speak at a conference who said, "The path to CEO is paved with P&L [profit and loss statement] roles." I anticipate that boards will spend more time looking at where individuals from historically underrepresented groups are landing in the organization. Is it more in the non-revenue driving roles? While representation data may remain constant or improve, what are retention and promotion looking like? That's an important question to ask.

I'm encouraged to see that some companies are thinking about how the emerging multiplicity of modules of working such as hybrid, remote-only, or physical presence-only impacts equity. Are people getting promoted more if they're in the office more? Are they not getting the attention or even the visibility they need using virtual platforms because they're remote? That's a whole other type of measurement of social performance and impact that is just getting started, the future of work.

Pagitsas: What is the number one issue you see organizations tackling around environmental issues and data?

DuBois: They are tackling measurements of their emissions with an eye to achieving net zero. Those methodologies vary quite widely, as well. In a recent article in *The Wall Street Journal*, the headline made me think it might be targeted toward questioning the veracity of emissions disclosure, but it was actually more exploratory. The writer explained how frequently companies are disclosing refinements or evolutions in their methodologies as knowledge improves. [Jean Eagelsham, "Companies Are Targeting Their Carbon Emissions, but the Data Can Be Tricky," September 3, 2021.] So, I don't know that the *E* is perfectly done quantitatively either. [laughs]

Pagitsas: Where are most companies on their sustainability or ESG strategy today?

DuBois: I would say at every age and every stage. The maturity curves are all over the place. And it's not just companies. It also varies across geographic regions and regulatory regimes. All those maturity curves start to cross over each other. Interestingly, any time a company is changing via acquisition or divestiture, going private, going public, splitting, or spinning, the business model changes. Therefore, the ESG strategy also needs to change to reflect that new business. We now see companies beginning to prepare their ESG goals, strategy, and profile in advance of going to the public markets, which is a newer trend. After they go public, they do have a bit of a grace period of twelve to eighteen months to put together their ESG strategy, but we are seeing it shift earlier.

Pagitsas: Are investors' expectations of an ESG strategy earlier if the company has a preexisting product or business lines aligned to ESG principles?

DuBois: Not necessarily. It's not always the "inherently ESG" companies that are moving faster at the enterprise level. This is an incremental process for most. You can't do it all at one time because nobody wants to get ahead of their skis. Honesty regarding your roadmap is the best policy. [laughs]

Pagitsas: This certainly emphasizes the need for strong governance so that only correct and accurate data is shared with investors and the public. What are your thoughts on governance? How can a company use strong governance to achieve their ESG goals?

DuBois: The lawyer side of me comes in on governance and accountability. You must cascade the ESG strategy throughout the organization, just like with any other business initiative, so board oversight is essential. While you'll also often see a C-suite member acting as a champion or holding ultimate responsibility for execution, that can vary across companies. As a C-suite leader or board member, you need to say, "If we have DEI goals that involve expanding the percentage of employees from historically underrepresented groups at a particular level of the company, what tactics are management deploying? For example, what does talent acquisition need to do on pipeline development? What learning and development are required to perform well on retention as representation numbers stabilize or increase? Do we need to require diverse slates as part of succession planning?" You must assign accountability to those who have the expertise to devise and implement the tactics.

This focus on accountability has clearly increased since the onset of COVID-19. The mechanisms for oversight and implementation are becoming a bigger part of the strategic planning process for many companies.

Pagitsas: What common roadblocks can companies remove to move forward with creating accountability?

DuBois: I've seen organizations that are starting out question whether they can make inroads without building a large ESG office. In my experience, however, you don't need a big office. ESG should be embedded in every function and business unit, so that the right people with the right expertise are executing the work.

I view the ESG role as an "up periscope" role. There are a few components that the ESG role may own—strategy development, data collection and reporting, and management of governance mechanisms and processes like regular C-suite and board reviews. ESG might be housed centrally, but it's a heavily networked role.

One practical resourcing solution could involve placing a high-potential early-career employee in an ESG project management role for a few years under the supervision of an executive sponsor. This would give them a 360-degree view of the company and send them back into the business with an ESG mindset. If more resources become available, they might consider investing in a data scientist to collect and interpret the data, as well as a technical ESG reporting expert who can collaborate with communications, investor relations, and other outward-facing functions to report on progress and tell the company's story.

Pagitsas: So, the ESG team may be a small team responsible for strategy development and infrastructure management, but who is responsible for executing the work to achieve the company's ESG goals?

DuBois: It depends on the goal. For example, for a manufacturer to reduce water use, operational expertise must be brought to bear. Similarly, a DEI goal does not reside solely with the chief DEI officer but perhaps also with an executive compensation expert who can help develop metrics to embed into executives' performance plans. Ultimately, there needs to be an appropriate network of people to implement the tactics. It's nitty-gritty when it comes to organizational design, staffing, and accountability mapping.

Pagitsas: What changes should any business anticipate with the increased focus on ESG?

DuBois: ESG can no longer be a buzzword used on marketing materials or a check the box exercise. Strong ESG programs pave the way for more sustainable businesses that can unlock greater long-term value, generate enhanced returns for investors, and positively impact society. Quite simply, an ESG strategy won't have legs without full and complete buy-in and integration across an organization.

More broadly, I see a couple of big trends. One relates to the ESG data environment. What I mean by that is that when ESG reporting began, it was primarily a look-back with organizations saying, "Here's the data from last year." Now I think you're seeing them talk more about their strategy for the future, which means that ESG is becoming *the* strategy.

On a related note, a PIMCO study of 10,000 companies globally showed that from 2005 to 2018, zero to one percent of earnings calls referenced ESG issues. From 2018 to 2019, that number increased to five percent. Post-COVID-19, it's at nineteen percent to twenty percent. [PIMCO, Erin Browne and Geraldine Sundstrom, "Mid-Cycle Investing: Time to Get Selective," September 14, 2021.]

In a recent article, FactSet also published data on the record number of references to ESG factors in earnings calls for the S&P 500. [John Butters, "Record-High Number of S&P 500 Companies Discussion 'ESG' on Q2 Earnings Calls," September 10, 2021.] These studies suggest that more strategic, future-facing information is becoming available and will continue to become a greater part of where the market is headed.

Within asset management, the path to ESG originated with socially responsible investing [SRI], which is primarily associated with the expression of a particular investor's personal or institutional values. Then came a shift from "values" to "value," reflecting a risk-return objective that can fit under the broad umbrella of ESG integration. It's possible that we will see values coming back in, potentially as a component of the total analysis of a company's potential to create value for stakeholders. This is a different lens for both companies and investors.

One expression of this development could involve investors digging more deeply into corporate culture. Boards are very focused on corporate culture right now, particularly in the wake of the pandemic. In addition, the $30 trillion generational transfer of assets to millennials may create large-scale change, depending on how they ultimately decide to invest. The term "impact investing" is more prominent in Europe right now, but the twin objectives of creating positive change while earning a desired rate of return are coming into the total examination. It's almost like values are experiencing a resurgence in a different context.

Pagitsas: Why is values integration coming back? What's driving that?

DuBois: Asset owners who hire asset managers are asking about impact. The asset managers are fiduciaries to their clients, so they're answering the call. Maybe it's driven by millennials and beyond. My daughter is Gen Z, so she's going to do everything herself. That's how they think. They may want to effect social or environmental change with their dollars because they want to do it themselves.

I do think that COVID-19, combined with the renewed quest for equity and social and racial justice, have made a lot of people reassess their objectives. A meaningful subset of investors may be doing a reset as a result. I don't know if it will last or not, but I think it's pretty prominent right now. Anecdotally, I'm hearing that asset owners want to talk about values. And Edelman's *Institutional Investors* research indicates that a majority of US investors are more likely to invest in companies that take a stand on a social issue, with access to health care and climate change in the lead.

Pagitsas: You have companies as legal entities being asked to express values, which are usually associated with humans as individuals. Is there a risk, Heidi, of companies as an institution, as a legal entity, being inauthentic?

DuBois: Yes. Nearly all US investors included in Edelman's most recent research say that they must have trust in a company's mission or purpose before recommending an investment.

Pagitsas: How can a company preserve its authenticity and demonstrate its values through its employees and its products and services? It's a very existential question, I know. [laughs]

DuBois: Law school was my introduction to the idea of the corporation. What has stuck with me since then is that it's really a social contract between the government and the corporate entity, which receives a license to operate so that it can provide goods and services that we can't produce on our own or that government isn't as well suited to deliver. We as citizens naturally have expectations for performance. If we're having a global discussion about the purpose of business, as opposed to the purpose of a particular corporation, we keep coming back to the question, what do we want business to do? And once we know what we want it to do, what is the scorecard for measuring its performance? A company should probably reevaluate its aims and priorities every three to five years.

So, let's talk about action communications. Act first, talk later. We see employees looking to their employers as the most trusted source of information about what's going on in the world. We see a high percentage of employees expecting their leadership to speak out on pressing social issues. In Edelman's research regarding the "belief-driven" employee, one finding really stood out. Six in ten respondents said they choose their employers based on beliefs and refuse to work at a company whose values they disagree with. In the current environment, employees are core to the multi-stakeholder model. [Richard Edelman, "The Belief-Driven Employee," Edelman Insights, August 31, 2021.]

Institutional investors continue to sharpen their focus on ESG as a critical trust factor. Edelman's *Institutional Investors* report found that many investors believe companies frequently exaggerate their ESG progress in financial disclosures, and most investors don't believe many companies will achieve their stated ESG goals. So, we still have some gaps to close as the ESG landscape continues to evolve.

Pagitsas: Coming back to your professional experience, what was your journey to your current role as head of ESG for AEA?

DuBois: I started my career in 1999 as a merger and acquisitions lawyer focusing on executive compensation. From there, I worked on public company disclosure and proxy issues. My first ESG issue was the explosion of executive compensation disclosure after the Enron debacle. I then started to work with boards on the governance and compensation front at several US publicly traded companies.

As I progressed into what is now called ESG, I found that my legal training had prepared me well for the diligence required and the reality of implementing an ESG goal. That background and its rigor made me tenacious on the ESG front.

Soon after I joined PepsiCo in 2011, the company appointed its first chief human rights officer. In addition to my governance and compensation roles within legal, I supported the development and implementation of a human rights program. That role marked my formal transition to a full-time ESG professional—a very big moment for me.

My next role was at BNY Mellon, where I served as the ESG "periscope" for the internal functions and business units around data collection and a node of collaboration with communications and investor relations. In addition, I was responsible for ESG reporting and engagement with non-governmental organizations. After ten years in corporate, I decided to return to an advisory role because, as a lawyer, I really enjoyed helping other people out. So, I joined Edelman to lead their advisory practice. At AEA, I partner with our portfolio companies to build and execute ESG strategies intended to create long-term value for our investors.

Pagitsas: What professional and personal experiences connect to you to the environmental, social, and governance issues you work on? How do they influence your leadership approach?

DuBois: The lawyer in me is the G, or *governance*, part of ESG. The S, or *social*, part is my work in human rights. And the E, or *environmental*, comes straight from the ocean. I love and respect the sea.

I grew up as a competitive sailor, racing boats with my family. In sailing, it's the mission first: everything should contribute to the mission. If an individual isn't focusing on the mission in sailing, you're down more than one crew member. Not only is that person not pursuing the mission, but others on the team must make up for it. I feel great when I'm part of a team that is operating as a seamless unit.

Pagitsas: Why are human rights important to you?

DuBois: When I first learned about the United Nations' Guiding Principles on Business and Human Rights while working at PepsiCo, I realized that they were the blueprint for the person I wanted to be. This framework captured how I want to treat other people in the world. For example, it helped change my approach to business meetings where perhaps the meeting would be contentious, or maybe there would be many opposing points of view or someone participating whom I might have to work harder to relate to. It changed the way I interacted with colleagues because I would remind myself that every person in the room is endowed with fundamental human rights. That's how they should all be treated by me. It did change my approach to interaction with others in the workplace and beyond. And the *Universal Declaration of Human Rights*, with Eleanor Roosevelt as its driving force, is always an aspiration.

Pagitsas: Heidi, what drives you every day?

DuBois: Every day, I have the opportunity to see businesses in a new way. It's the anticipation of what's to be discovered that drives me. For example, in decades past, our ability to model potential shifts in temperature rise, what that might do to agricultural production, and what kind of supply chain impacts might result was much more limited. It's so exciting to uncover the potential for running businesses better. That's what really has me thinking about ESG all the time. I also like it because it's complex. These issues are often intertwined, and tradeoffs often have to be considered.

Pagitsas: With ESG, there is no one-size-fits-all answer to the issues. Every answer is in some ways new and crafted for a unique situation. Is that correct?

DuBois: Yes. With this work, there's a certain degree of riding the volatility that you need to live with. I get satisfaction out of helping other people feel more confident about being able to do that.

DuBois: At least that's my goal. I don't know if I will succeed. [laughs]

Pagitsas: What is your favorite book and why?

DuBois: Oh, that's easy. My favorite book is *Death Comes for the Archbishop* by Willa Cather [(Alfred A. Knopf, 1927)]. I think it is because of its contemplative appreciation of the landscape and the people who move through it. I read the book for the first time in seventh grade, and I wrote my first research paper about it then. There's a passage that describes how the Diné people, who used to be called the Navajo, move through the land. During COVID-19, I would visit our roof and see that the sky remained huge, that the grass was still growing, and that flowers continued to bloom. It's good to feel like a small part of a much greater world. That's what that book has always done for me.

The only thing that I learned much later was that Willa Cather quietly lived in her own way. She didn't define herself by prevailing conventions of femininity— she just was. I admire people who can live calmly as their true selves—an aspiration I also try to pursue.

Marisa Buchanan

Managing Director, Global Head of Sustainability
JPMorgan Chase & Co.

Marisa Buchanan is managing director and global head of sustainability at JPMorgan Chase & Co., an American multinational investment bank and financial services holding company. JPMorgan Chase is the largest bank in the United States and the fourth-largest bank in the world in terms of total assets, which were $3.4 trillion as of year-end 2020.

Marisa has over 20 years of experience working at the nexus of business and the environment. Marisa is responsible for developing and implementing the firm's sustainability strategy across its financing activities and operations. In this internally and externally facing role, she plays an integral role in JPMorgan Chase's efforts to drive sustainability initiatives and commitments forward, including the firm's Paris Agreement-aligned financing commitment and $2.5 trillion sustainable development target. She has also built the firm's knowledge and leadership on climate change and carbon disclosure, and leads the firm's efforts on ESG reporting, stakeholder and policy engagement, and sustainability grantmaking.

© Chrissa Pagitsas 2022
C. Pagitsas, *Chief Sustainability Officers At Work*,
https://doi.org/10.1007/978-1-4842-7866-6_21

Prior to joining JPMorgan Chase in 2012, Marisa was an analyst at Bloomberg, where her research focused on the impacts of US policy and regulation on the electric power and oil and gas sectors. She developed methane reduction projects in the mining and oil and gas industries with Verdeo Group. In addition to identifying and evaluating project development opportunities, she managed activities related to the certification of carbon assets and led engagement on US and international climate and energy policy. She also developed renewable energy and methane reduction projects with Econergy International, an independent power producer with assets in Latin America and the United States, which GDF Suez acquired. She began her career with the Surdna Foundation's environmental team.

Marisa serves on the board of the Environmental Law Institute. She was recognized in 2021 as one of the Environment + Energy Leader 100 honorees for her contributions to the advancement of sustainable solutions.

Chrissa Pagitsas: When did you first become aware of the impact of environmental and social issues and their relationship to businesses?

Marisa Buchanan: I spent my young adult years in the Bay Area in California in the early to mid-nineties. It was a period of immense growth in the region. I saw a whole host of challenges, such as affordable housing constraints and huge issues around traffic congestion. The summer before my senior year in college, I interned for a trade association representing big businesses in the Bay Area, now called the Silicon Valley Leadership Group. The organization was engaging on "quality of life" public policy issues in the state such as housing, public transportation, and land use that matter to businesses trying to attract and retain the best talent and build successful organizations. It was the first time I got to see how public policy matters to the business community and how good policy can make a real difference in local communities.

Shortly after I graduated from college, I got a job at the Surdna Foundation in New York City. I worked in the environment program, where we made grants to non-profit organizations working to advance issues such as sustainable forestry and renewable energy, but also smart growth and market-based solutions to climate change—relatively new concepts at the time. That experience helped me understand the deep connection that exists between environmental and social issues. It also sparked my interest in working from within the business community to tackle challenges like climate change.

My time at Surdna also sparked a deep interest in energy and climate change. I find energy and the role it plays in the global economy to be fascinating. Energy is instrumental in promoting global economic development growth and social advancement, but the world's historic reliance on fossil fuels has come with a cost. The challenge is figuring out how to provide energy for people's needs while doing so affordably and with a much lower environmental impact. It's important to remember, though, that there's no energy source without trade-offs. For example, we need steel to build wind turbines and rare earth minerals to build solar panels and batteries. Geopolitical issues come into play, too. The scale of the challenge is immense, but we need to keep making progress.

Pagitsas: Was there a particular experience in your career that solidified the interconnectedness between social, environmental, and business issues for you?

Buchanan: After I worked at Surdna, I went to graduate school for my master's degree, where I studied environmental and energy policy, economics, and finance. When I graduated, I went to work for a renewable energy company that was developing projects under the Clean Development Mechanism [CDM], set up through the Kyoto Protocol. This was the first large-scale compliance offset market, where entities in developed countries could essentially invest in emissions-reducing projects in developing countries. The company was developing wind and run-of-river hydro projects in Latin America for the CDM market and had a small team focused on methane mitigation projects in the United States, where the voluntary carbon market was beginning to grow.

In that experience, I was able to get my hands into project financials and the mechanics of what it takes to actually develop a project. One lesson that I learned is that something can be good for the environment, but the economics must work for businesses and investors. Economic sustainability and environmental sustainability have to go hand in hand. But it's not all about economics. Sometimes the economics can work, but there are other barriers that prevent projects from being developed.

Pagitsas: If it doesn't make financial sense, what should companies do to address environmental challenges? Is there a spectrum of responses possible in these scenarios?

Buchanan: Most companies aren't running their businesses just for today. They're constantly thinking about how to build and invest for the future. And just because something doesn't make financial sense today doesn't mean it won't in the future. Bringing costs down is a critical piece of developing and commercializing new technologies and solutions. It takes investment and time, but it can pay off significantly in the long run.

Another aspect is understanding that the cost and benefit of taking certain actions are not necessarily all financial, nor can they be easily translated into financial metrics. Companies must ask whether there are reputational, customer, or employee loyalty benefits to taking a particular action. Or if there is a long-term competitive advantage worth capitalizing on or a risk that we face by not taking an action that has an environmental benefit. And sometimes, a certain action can just be the right thing to do.

Pagitsas: What are other key takeaways you've had about integrating businesses and addressing environmental issues?

Buchanan: Between 2008 and 2010, I worked for a small startup working with coal mining companies and oil and gas companies to help them reduce fugitive methane emissions, which contribute to climate change. I spent a lot of time in West Virginia, Pennsylvania, Colorado, Wyoming, and Oklahoma. This was during a time where the business community was starting to engage more deeply on climate, but also when there was still a fair amount of skepticism within some industries and political circles that climate change was a real thing. Our pitch was to encourage companies to reduce fugitive methane without making climate change the reason they should do it. In this case, efficiency mattered hugely to our clients. They wanted to operate responsibly and not waste a valuable resource, especially one that could be used and sold. We were able to get some great projects developed because of that.

Pagitsas: So, you talked money, but you didn't talk environment with them?

Buchanan: It's not that we didn't talk about climate. What we weren't saying is that climate is the number one reason you should spend money on new technologies to capture methane. For me, that experience underscored that when it comes to getting something done, you have to understand what matters to the person or company you're trying to work with. What are their interests and motivations? What do they care about? Listening and understanding are essential.

It also underscored for me that most of us have far more in common than we recognize. But, too often, people focus on differences and don't think about where common interests intersect and how to focus on driving something forward, which is good for both sides.

Pagitsas: Now you're at JPMorgan Chase. What is the scope of your role?

Buchanan: As the head of JPMorgan Chase's sustainability team, my team and I are responsible for developing the firm's sustainability strategy across its financing activities and operations. We also oversee ESG reporting, stakeholder and policy engagement, and sustainability grantmaking. Much of what we've been focusing on recently is helping to build the sustainability expertise and capacity across the organization, especially in our lines of business that serve our clients and customers on a daily basis.

Pagitsas: How does JPMorgan Chase define sustainable finance?

Buchanan: It's about helping our clients raise capital to achieve their sustainability objectives. It's also about helping our clients invest their capital in ways that align with their values or meet certain sustainability objectives. In addition to financing and investing, we also provide our clients and customers with advice and research. We're focused on building our own organizational expertise and capacity across all these functions. That's how finance becomes more sustainable. It must be integrated into, not separate from, what we offer to our clients and customers.

Pagitsas: The Center for Carbon Transition [CCT] is for any JPMorgan Chase client seeking to finance commitments to meet the goals of the Paris Agreement, correct?

Buchanan: The CCT is an example of the kind of capabilities we're building to better serve our clients. We announced the creation of the CCT when we made our Paris Agreement-aligned financing commitment in October 2020. We know a growing number of our clients across different industries need financing to be able to invest in and transition their businesses over time. That could include loans but also capital raising in the private or public markets. But they'll also need things like advice and access to research.

The CCT is there to help make sure that our clients are getting access to the full suite of capabilities that JPMorgan Chase has to offer, so they leverage the expertise and offerings of many other teams around the firm—whether that's coverage banking, debt, and equity capital markets, commodities, sales and trading, and others.

In 2021, we also created a Green Economy team in our Commercial Banking business to provide deeper coverage of small and medium-sized companies that produce environmentally friendly goods and services or focus on things like environmental conservation. We're very focused on making sure we have the right people and enough people in the client coverage roles and other product and services roles so we can meet the changing needs of our clients and customers.

Pagitsas: What is the Paris Agreement?

Buchanan: The Paris Agreement is a treaty supported by over 190 countries to reduce global greenhouse gas emissions to limit average temperatures from rising more than two degrees Celsius above pre-industrial levels and, ideally, no more than one and a half degrees Celsius. Individual countries have come up with their own climate plans, called Nationally Determined Contributions, to reduce emissions to help achieve the goals of Paris. So, the global "roadmap" to Paris is essentially the sum of all the plans that countries have developed.

The challenge is that the measures taken by governments to date aren't enough to reduce emissions to a level needed to reach the Paris goals. It's a gap between ambition and reality. This is one of the reasons why there's so much pressure on the private sector to fill that void.

Pagitsas: What is JPMorgan Chase's commitment to the Paris Agreement?

Buchanan: JPMorgan Chase has long supported the need for thoughtful energy and climate policies. In 2017, when the United States announced plans to withdraw from the Paris Agreement, we said publicly that we thought the United States should stay at the table. And we were supportive of the United States rejoining in 2021.

For us, the biggest role we can play on climate is through the work we do with our clients and customers. On the one hand, that means doing more to finance green technologies and helping companies across the world access capital so they can invest in measures to reduce emissions within their operations or create new products and services that enable their customers to do so. This is why we set a target in 2021 to finance and facilitate more than $2.5 trillion through the end of 2030 to advance long-term solutions that address climate change and contribute to sustainable development. This includes a $1 trillion subtarget specifically for green initiatives.

On the other hand, it also means working with companies that are operating responsibly and thinking about climate as a strategic risk and opportunity. In October 2020, we were among the first banks to announce a Paris Agreement–aligned financing commitment. We've since published emissions targets for our oil and gas, electric power, and automotive financing portfolios. This means that, by 2030, we're going to reduce the carbon intensity of those financing portfolios to a level that aligns with the longer-term goals of Paris. In the fall of 2021, we also joined the Net-Zero Banking Alliance and have plans to establish targets for other industries over time. Our asset management business is also a member of the Net Zero Asset Managers initiative.

Pagitsas: Why is it important to help companies align with the goals of the Paris Agreement?

Buchanan: From my prior experience, I know that if you want companies to transition, you have to help them. The alternative is walking away. In my view, that's the wrong approach. The reality is that the world needs a lot of energy today and is going to need a lot more of it in the future. The business community is constantly innovating, but the path to net zero is going to require a massive transition in the way that the world both sources and consumes energy. We need to support that and help accelerate it.

Pagitsas: What are the critiques you hear of this transition strategy? How do you address the critiques?

Buchanan: I think it's important to look at the realities of our global energy system and economy. Too often, the conversation around addressing climate change is about reducing fossil fuels, and specifically, supply. But reducing supply without also reducing demand is not a recipe for success. People rely on those energy sources today. Things like wind and solar have grown exponentially, but the world uses fossil fuels for transportation whether it's trucking, shipping, or aviation and to produce things like steel and cement that are used in industry and real estate. The world also uses oil and natural gas to produce a huge range of chemicals, plastics, and products, including all sorts of things that many people use daily.

The transition will be hugely complex. In addition to reducing greenhouse gas emissions, issues like energy cost and access, the availability or lack of resources in certain areas, and geopolitical considerations all have to be considered. I'm a big believer that, for climate solutions to be successful, they need to work for people and communities.

Pagitsas: How is the sustainability strategy you've helped shape being received internally at JPMorgan Chase?

Buchanan: When I arrived at JPMorgan Chase almost ten years ago, there were some who asked, "Why do we have a sustainability team? What is your job?" It was curiosity, not hostility. Things are so different today. I always say, "Organizational change occurs due to a confluence of internal and external factors." In this case, it's been growing societal recognition of climate change as a threat, advances in technology that are making many low- and zero-carbon technologies cost-competitive, policy and regulatory changes, and a desire—by many of our shareholders, clients, and customers—for us to be doing more. Our firm's leadership has also acknowledged these trends and understood how important it was for us to lead.

Pagitsas: What else is a factor in the global focus on sustainability?

Buchanan: Media—especially social media—has done a huge amount to enable people and activists on the ground to document issues and publicize them, and to hold companies accountable. That's something that doesn't get as much attention as it should and deserves a lot of credit for opening many people's eyes to things like environmental degradation and pollution.

I know that has positioned activists and others to really work with the investor community to help push companies. You see that and the extent to which so many institutional investors are engaged today around environmental and social issues. It used to be just about governance. But now it's so much more than that. If you're a CEO or a CFO now and want to raise capital, investors want to see that you're acting responsibly. They want to understand the impact that you're having on communities in the environment. That's when the CEO or CFO says, "We've got to have a sustainability strategy." But it's not only a strategy. So much of it is finding a way to tell your story about what you're doing and why it makes you a better business. As a company, you must tell that story to so many different audiences today.

Today, the global flow of information is pretty fluid and real time. However, in my view, one downside of the prevalence of social media and the twenty-four-hour news cycle—which has shortened our collective attention spans and created a need for attention-grabbing headlines—is that it's not been conducive to promoting a rational conversation about the complexities of the challenges we face on energy and climate change. The challenges and solutions aren't black and white, but that's often how they're presented.

Pagitsas: Lastly, as a seasoned sustainability leader, what advice would you give someone just starting their sustainability career?

Buchanan: I was recruited here by Matt Arnold, who was my boss for nine years. Years prior, he had worked for an organization that was one of our grantees at the Surdna Foundation. I've always followed people I know and respect and who inspire me. That has been a theme throughout my career. I always tell people looking for advice, "Find somebody you want to work for." Forget titles, so on and so forth because, in the sustainability field, no job description is ever fixed. Your job is basically what you want to create it to be. And it changes every year as you make progress. That and learning how to be an effective influencer are really important.

Cynthia Curtis

Senior Vice President, Sustainability Stakeholder Engagement
Jones Lang LaSalle

Cynthia Curtis is senior vice president of sustainability stakeholder engagement for Jones Lang LaSalle (JLL), responsible for embedding JLL's sustainability program throughout all business lines and driving meaningful environmental, social, and financial impact with and through JLL's clients. JLL is a global commercial real estate services company with offices in more than 80 countries and over $16 billion in revenue in 2020. Cynthia is driving JLL to deliver on its net-zero carbon target of reducing scope 3 emissions from the properties it manages by 2040. Additionally, she ensures JLL's investors have a complete understanding of JLL's competencies, goals, and impacts through reporting in accordance with the Task Force on Climate-related Financial Disclosures (TCFD) standards.

Previously, Cynthia has worked in the public, private and non-profit sectors, including Ceres and CA Technologies, where she served as vice president and chief sustainability officer.

Cynthia represents JLL on the Corporate Advisory Board of the World Green Building Council (World GBC). She is also on the board of directors for Greenbacker Renewable Energy Company, which is dedicated to investing in projects and managing capital for its public shareholders and institutional investors in the sustainable infrastructure space.

© Chrissa Pagitsas 2022
C. Pagitsas, *Chief Sustainability Officers At Work*,
https://doi.org/10.1007/978-1-4842-7866-6_22

Chrissa Pagitsas: What was your professional journey to your current role as senior vice president of sustainability stakeholder engagement at JLL?

Cynthia Curtis: My career, I would say, has been a rather circuitous route. The bulk of my career was in marketing, and I still consider myself a marketer. I spent the first five years on Capitol Hill in Washington, DC, and then got my MBA to focus more on international business, with an emphasis on marketing. In 1995, I started working at tech companies—Unisys, followed by EMC. I spent a good part of my career running services marketing globally for these firms.

At EMC in the early 2000s, I started hearing from clients that our technology consumed too much energy. Although my boss at the time said it was an engineering issue, I pushed back as I felt there was something more here to explore, like getting greater standardization. I was thinking ahead to policies that may impact EMC and the energy consumption of its equipment, how to get ahead of that, and how to influence it. That was my background coming from Capitol Hill. But I didn't sell it very well, and it ended there. This was before sustainability was as it's known now. Fast-forward and EMC, now Dell, is taking some very meaningful steps. In fact, one of my good friends, Kathrin Winkler, got them going on their journey.

Then in 2006, I watched the movie *An Inconvenient Truth*, which spurred me to ask, "How can I *not* be doing something about these climate issues?" I felt very strongly that if business didn't see a benefit from shifting focus to make some environmentally sustainable modifications and investments, then no lasting change would happen. It was my catalyst and made me determined to leverage my skills and experience to make a difference.

Pagitsas: How has your passion for sustainability manifested itself in your personal life?

Curtis: You could say that I was an early adopter green buildings. I built my own house certified to the US Green Building Council LEED [Leadership in Energy and Environmental Design] Gold standard. At the time, it was very novel in the residential space, and my house was the first in our town to achieve LEED certification. If I was going to build, I decided it needed to have a minimal environmental impact. My architect said, "Well then, we should go for LEED certification." What a tremendous learning experience it was for all of us—me, my architect, and my builder. Aside from all the documentation requirements, every decision had multiple permutations in terms of materials to choose from. Because everything was so new at the time, I paid a premium. In marketing parlance, it was a "first mover" disadvantage! That said, I'd do it again in a heartbeat.

Pagitsas: In what ways did this personal experience translate into your professional career?

Curtis: Absolutely! It was a fillip to my focus on sustainability at CA, then at Ceres and now at JLL. It helped me decide to course-change my career to focus on sustainability. And it gave me some credibility because it was tangible evidence of my commitment. I had an opportunity come my way with CA Technologies to lead sustainability communications and strategy. I thought to myself, "I can do that all day long." But on my third day, the guy who hired me and who was the only other person with "sustainability" in their title at the company was let go. Consequently, my learning curve went from 90 degrees to 180. It was pretty steep, but it also allowed me to create and lay the foundation of what I thought we needed to do.

After CA, I spent time with Ceres leading up to the 2015 UN Framework Convention on Climate Change [UNFCCC], also known as the COP or the Conference of the Parties, in Paris. One of the things that I was most excited about and thankful for was the opportunity to develop a statement by the largest US banks in support of action to address climate change. The banks had never issued a joint statement on anything before, let alone climate change. They were motivated to issue this statement at that time because they had begun to get pressure from their stakeholders to act on climate change. It was quite an exercise in negotiation and diplomacy, but we ended up getting it out.

After that, a guy named Michael Jordan from JLL, who I had hired to help me when I was at CA Technologies as part of a consulting project, called me up and asked, "Do you want to come over here to JLL? You can do sustainability consulting with our clients." I thought, "I don't know if I really want to do that, but we'll give it a go."

So, I joined JLL in 2015, working initially with clients on their sustainability strategy but realized that we weren't really walking the talk as much as I knew we could. Hence, I shifted to focus half-time on JLL's internal program before moving 100 percent over to the corporate sustainability team to help develop and drive JLL's strategy.

Pagitsas: What is the scope of your role?

Curtis: My scope is to drive the sustainability strategy, work with our clients and our client-facing folks, reach out to our investors, partner with the investor relations [IR] team, and represent JLL at member organizations like Ceres and World Green Building Council.

Pagitsas: Why is stakeholder engagement a critical component of JLL's sustainability strategy to identify and around which to create a strategy?

Curtis: It's an essential component of any sustainability effort. More than any other function sustainability requires collaboration—an overused but under-executed term. Fundamentally, the role is to inform our strategy development and execution. To do that, you must have input and dialogue and hold a mirror up to yourself in terms of ambition and performance. Otherwise, you risk, as my father would say, just "gazing at your navel."

Pagitsas: Coming back to your day-to-day work, it sounds like your role includes consulting internally to meet JLL's own sustainability goals as if they were a client. Is that accurate?

Curtis: It's a part of it. But frankly, any sustainability leader in any company is, to some degree, an internal consultant and an evangelist within the organization. The rewarding thing for me is that over the time that I've been at JLL, the requirement for me to be that internal consultant is diminishing because we are succeeding at embedding sustainability more broadly across the company. Our senior management team truly recognizes the real value and the importance of bringing sustainability more to the core of our strategy.

Pagitsas: Was there a particular moment when it hit you that you didn't need to be the evangelist?

Curtis: The role of an evangelist never goes away. And I wouldn't want it to! Every organization needs a champion. But, hopefully, it evolves. For me, the most demonstrable evidence of that progress was with our IR team. Included in my remit is to engage with our shareholder investor community about what JLL is doing in sustainability, how to better tell our story, how to help investors understand it, and thereby integrate it into their consideration of our value. Part of increasing that engagement was getting more sustainability content into our quarterly earnings calls.

Early in my tenure, I would draft key bullets on our sustainability program, goals, and ongoing activities, and send them off to the folks I was working with in IR for potential inclusion in the content. But I wasn't getting much traction. As with most companies, investors weren't raising the topic on our earnings calls, so the assumption was they weren't interested. I tried a different tack, "What if we have an ESG [environmental, social, and governance] focused webinar?" Initially, there was hesitation, but they agreed to give it a try. At the time, no one other than IR, our CFO, and CEO, spoke directly with our investors. So, I had to work hard to convince our leaders of the benefits to investors in hearing from the sustainability leaders. At first, the requirement was that my boss, Richard Batten, and I do a recorded webinar. My initial response was, "There's not much dialogue in that approach." It wasn't going to create the give and take with investors that we wanted. However, I understood their perspective. We'd never had an ESG-focused conversation before with them. Understandably, there was some trepidation about it. So, we taped a rehearsal, played it back, and improved it to the point where we could be live.

Pagitsas: How did the live webinar go with investors?

Curtis: It wasn't great because it was still felt a little scripted. But we had broken through, and people were positive about it. So, when we suggested doing a second live webinar, the response from IR was, "Okay, with practice." The result was a significantly improved, *much* more fluid, interesting conversation. Now fast-forward. IR regularly comes to me and says, "Okay, these are the points that we want to cover on our earnings call around sustainability. What are we missing? Anything else to add?"

Pagitsas: It seems the IR team is now comfortable with the ESG and sustainability strategy and language and is now ready to own it. What a great evolution! How long did it take for this change to happen?

Curtis: I'm going to say it was about eighteen months to two years. That was a significant shift. It takes time. We are in lockstep now. I just had a call this morning with someone from IR about our next webinar. We're planning it together, and it's going to be great.

Pagitsas: Let's talk about the people on the other side of the webinar, the investors. Were they equity or bond investors?

Curtis: Mostly equity. We worked with Goldman Sachs' GS SUSTAIN team. We wanted to not only educate our current investors but to go beyond that. One of my objectives was—and still is—to increase the collaboration, dialogue, and transparency with our existing investors and reach new investors. How can we expand the share of wallet with our existing investors, and how do we attract new investors? Generally speaking, when you have an investor who leans toward ESG, they tend to hold their investments longer—although not always. Part of our strategy is to access that longer-term capital.

Pagitsas: With the webinar, were you able to achieve this dual strategy of expanding JLL's existing investors wallet but also tap new investors?

Curtis: It's hard to draw that direct line, but we did see some additional investors that came in afterward. You know, there's no single event that brings in a new investor or client.

Pagitsas: I agree. Investor education is a long-term investment. It's not a "one and done."

Curtis: Yes, that's right. Within thirty days after the webinar, we gained a couple of new investors.

Pagitsas: Any other developments since the investor webinar affecting the company financially?

Curtis: Something that is a bit on the leading-edge side is that we amended our credit agreement to include links to ESG metrics. As a result, we're getting favorable rates if we exceed our targets and less so if we miss. This outcome of our ESG strategy is very real life. It's not a gazillion dollars, but it's something, and it shows that there is recognition on Wall Street that these investments are fiscally responsible.

Pagitsas: Speaking of the long-term, JLL announced in 2021 a major commitment to achieving net-zero emissions by 2040. What exactly is the scope of the commitment?

Curtis: In a word, *big*! Our net-zero carbon commitment is extremely ambitious. We've agreed to an absolute ninety-five percent reduction in emissions across all our operations, what's known as scopes 1 and 2, and across the properties we manage on behalf of our clients, known as scope 3, by 2040, with no more than five percent in offsets. It's daunting—particularly because around ninety-five percent of our emissions are client-based. But what an opportunity!

Obviously, our net-zero commitment isn't one of those things that come out of the blue. I'd been leading calls with our stakeholders for quite some time to understand what they were expecting around future environmental and social commitments. The team shared our proposed targets while knowing that they were not as ambitious as perhaps we needed them to be. Initially, we were not targeting a net-zero goal. As if by request and on command, our stakeholders said, "If your tagline is to 'achieve high ambitions' and if this is what you're considering ambitious, well, we think you can do better."

Pagitsas: Who were those stakeholders?

Curtis: The stakeholders included our clients, investors, NGOs [non-governmental organizations], academia, and other organizations focusing on organizational diversity. This exercise was about holding up a mirror to ourselves. It's something that we do with Ceres regularly. The stakeholders pushed back and said, "You've got some opportunity here, and you're not taking it. These goals aren't cutting it."

To be frank, even outside of the formal stakeholder calls, we were getting questions from some of our key investors and clients, such as, "Are you going to sign up for a net-zero target? Are you looking at The Climate Pledge that Amazon and Global Optimism have had put together? What are you going to do?" They were pressuring us to stretch.

Pagitsas: How did you respond to the stakeholders' critiques and questions?

Curtis: It was pivotal. We went back to the drawing board and did our homework to figure out the pathways and costs to make a net-zero goal happen. We went through the detailed process of determining what was necessary. What does it look like? How can we possibly get there with at least some specificity? Can we see a path today, knowing it will evolve? Essentially, we drafted a business case and took it to our leadership team.

Once we had their approval, we got our goal approved by the Science Based Targets initiative [SBTi], which laid the groundwork for us to come forward with our net-zero 2040 target in May 2021. Our decision to seek target validation by SBTi was to add third-party credibility. They didn't just rubber-stamp it. They reviewed and challenged some of our assumptions. There was some back and forth. We also, to be honest, welcomed the additional scrutiny. Setting such an ambitious target for your company is scary! We wanted as many fingerprints on it as we could get. Our target was inclusive of all scopes but within certain parameters such as operational influence for our scope 3 client emissions.

Pagitsas: Why was setting a science-based target important?

Curtis: Essentially, if this is what the science says, how can you do anything but? The initial reaction was overwhelmingly one of pride. Our employees were psyched at the announcement. But there was also some skepticism, not about the why but the how.

Pagitsas: Why is a net-zero goal important in the building sector and to JLL? I could argue that JLL doesn't own buildings. Its clients do.

Curtis: The built sector is responsible for approximately forty percent of the world's greenhouse gas emissions. We have a responsibility, an obligation, and a desire to lead the actions associated with driving that kind of systemic change. To do nothing as a company, you risk falling behind. Historically, the real estate as an industry is fairly conservative. But JLL is pushing the envelope when it comes to sustainability and ESG. We want to play a leadership role.

Pagitsas: What challenges do you anticipate as JLL tackles the net-zero commitment?

Curtis: Our net-zero 2040 goal stipulates that no more than five percent of the emissions reductions will be achieved through carbon offsets. Carbon offsets are a way of reducing your carbon emissions by investing in projects outside of your direct activities, such as renewable energy or methane capture from landfills.

That means we will be abating ninety-five percent of our emissions within scopes 1, 2, and 3, which is quite ambitious. Excluded from that are the products that we procure on behalf of our clients. Less than or equal to five percent offsets adds to the challenge for us, but we're committed to it.

Collectively we felt strongly that it couldn't just be a statement. There's enough uncertainty about the specifics of what's going to happen in another five years with the environment, the economy, and technologies. But we wanted to commit to a credible target today with a plan behind it.

Pagitsas: Who owns the plan to achieve net zero?

Curtis: This will absolutely be owned across the business. We want it to be ubiquitous so that different functions can see their roles in helping to drive achievement. Because it isn't just about what we do with our own workspaces. It is about what we do with our clients. It is about what we do with our supply chain. It touches all facets of the organization and value chain.

Pagitsas: Let's say that I'm a JLL client and highly hesitant, Cynthia, about setting a net-zero goal. But all my clients are telling me I must set a net-zero goal. What are the key reasons to set a net-zero goal?

Curtis: Aside from the growing evidence when looking out your window of how our world is changing, now, today, it's about risk and opportunity. We've delved into that analysis in our TCFD report, where we evaluated the potential impact of climate change to our business. The way that we as businesses have historically planned, forecasted, and measured was built on the perception and the understanding that our climate will be constant and very predictable. That is no longer the case. Actuarial tables are not as useful as they once were.

I see setting a net-zero goal is an example of good, classic enterprise risk management. In evaluating risk, you're looking at the best case, worst case, and most likely. You integrate that risk view into your planning and investments. And it's not just climate. The regulatory constructs are coming into play. It's smart to get ahead, or at least prepare for that inevitability.

There's also talent retention and attraction. I don't care what your business is. There are very few organizations that don't require smart, talented people. To energize and galvanize a workforce, you also need to give people something aspirational. You need to show that there's more than the quarterly bottom line, which is part and parcel of a good enterprise sustainability strategy. The *why* is that it's good, smart business. It sounds kind of simplistic, but in my mind, it's the foundation.

Pagitsas: As your fictitious client, I've still got cold feet, Cynthia. My company has been doing the same thing for the last one hundred years. What risks would you advise me to be aware of as I continue to debate setting a net-zero goal?

Curtis: Frankly, there's a risk of not doing it. The very real risk is that your asset will not be as valuable if you opt not to take these steps. It's what we call the "brown discount." In a worst-case scenario, you could be left with stranded assets. At JLL, we've researched occupancy rates and rental rates of commercial properties. We have correlated them with the actions that asset owners are taking around sustainability and greening of the property. The evidence shows that those assets are more valuable over time. That is one side of the equation.

Pagitsas: Isn't going net zero going to cost my fictitious company a lot of money?

Curtis: It doesn't have to. Done right, it will actually increase short-term operational efficiencies and the long-term value of the asset. It's what we at JLL call a "Return on Sustainability"—not just value creation but also value preservation.

Some investments, for example, have an immediate payback. One of them is Turntide, an innovative motor designed by a technology company that JLL has invested in. They refer to their motor as "the LED [light-emitting diode] lightbulb of motors." Motors aren't very sexy, but they are everywhere in a building. You swap out your fluorescent lightbulb for an LED lightbulb, and you immediately have a positive financial payback. The same thing holds true with this new technology—you have an immediate payback. A growing number of options do not necessarily make sustainable operations a more costly action or set of actions.

While all the answers aren't there, it's truly exciting to see some of the innovations that are growing out of companies saying, "We have to lower our company's emissions to zero. How are we going to do this? We need help." The investments, creativity, and brainpower going into these new innovations are exciting. This is where the opportunity comes in, capitalizing on the new markets that are being created. And I have to say it gives me hope.

Pagitsas: What if I challenged you further as this reluctant client and said, "Current regulations don't ask me about my company's ESG, sustainability, or greenhouse gas performance. That's another reason for my company to not change and not make a commitment to reducing emissions."

Curtis: I wouldn't have said this a year ago, but regulations are on the horizon. They're not down the street. They are around the corner. The SEC [US Securities and Exchange Commission] is in the process right now of drafting them in the United States. They already exist in the United Kingdom and parts of Europe. You don't want to get caught unprepared. What you want to be doing is helping shape what those regulations might be. The worst position to be in is to have them happen to you and not be prepared in any way to handle and to absorb them within your organization.

There's going to be a price on carbon. Whatever that price might be, it will happen. I think you can very rarely be so definitive, but it's going to happen. The best way to deal with any kind of carbon price is to reduce your carbon. Thinking again about your enterprise risk management approach, how do you best deal with that? You take that risk out of your product, out of your business.

Chapter 22 | Cynthia Curtis: *Senior Vice President, Sustainability Stakeholder Engagement, Jones Lang LaSalle*

Pagitsas: Thanks, Cynthia, for explaining to me, your fictitious client, the value proposition of going net zero. You've convinced me! Have you seen this same shift in conversations with real clients?

Curtis: Totally. This is now part of every client conversation, and it's on the shortlist of priorities. There's a growing appreciation of the relevance of sustainability to a business's success. In these conversations, you're not pushing boulders up the hill anymore. You're pushing on an open door.

Steve Waygood

Chief Responsible Investment Officer

Aviva Investors

Dr. Steve Waygood is the chief responsible ivestment officer at Aviva Investors. Aviva plc is a British multinational insurance company with more than 33 million customers and approximately $300 billion in assets under management as of January 2022. Steve joined Aviva in 2006, founding its Global Responsible Investment function and its Sustainable Finance Centre for Excellence.

Steve started his career at the World Wide Fund for Nature–UK in 1995, where he worked on its ethical and environmental investment policy. He subsequently worked for three buy-side fund managers. He has advised the Government of the United Kingdom, the European Commission, the Financial Stability Board, and the United Nations on the creation of sustainable capital markets. He also now serves as a trustee on the WWF–UK board.

Steve is actively engaged in international standards-setting and governing bodies. He co-founded the Corporate Human Rights Benchmark, the World Benchmarking Alliance, and the UN Sustainable Stock Exchanges Initiative. Steve is also a member of the Financial Stability Board Task Force on Climate-related Financial Disclosures (TCFD).

Steve has received the Leadership in Sustainability award from the Corporation of London and the British Chamber of Commerce. He has also received awards from Brummell, the International Chamber of Commerce, the City of London, the United

© Chrissa Pagitsas 2022
C. Pagitsas, *Chief Sustainability Officers At Work*,
https://doi.org/10.1007/978-1-4842-7866-6_23

Nations Foundation, Yale, and Harvard. He was highlighted in the Financial Times *in 2018 as one of the warriors of climate change among money managers.*

Chrissa Pagitsas: Let's begin by framing Aviva's services and the role it plays for many individuals.

Steve Waygood: Aviva is a global insurance firm that offers life and general insurance services to over thirty-three million customers across sixteen countries. Within the United Kingdom, Aviva is the leading life and pensioner provider with over eighteen million pensioners. Its asset management business is called Aviva Investors. In our role as investors, we look to integrate environmental, social, and governance [ESG] issues into our investment philosophy. ESG is also front and center of our business strategy, alongside a focus on enhanced customer outcomes and simplification. We are here to deliver the outcomes that matter most to today's investors.

For us, investing responsibly is not a fad. It is a core investment belief. As an active owner of capital, our scale and influence help us drive the change required to build a future our clients are able to and want to retire into.

Pagitsas: What is the scope of your role at Aviva Investors?

Waygood: As chief responsible investment officer, I advise our CEO and board on our overall ESG strategy and approach. I also lead our macro-stewardship work, which I define as systematically engaging with governments, regulators, policymakers, and even NGOs [non-governmental organizations] on material market failures in order to promote specific measures for their correction. I believe investors have a vital role to play in pushing for change on society's biggest issues, from climate change to diversity, from environmental degradation to human rights. By investing responsibly, including engaging with companies at the micro level and with governments and regulators at the macro level, we can help achieve inclusive economic growth, environmental protection, and social development. The latter part of this is a personal focus for me.

Pagitsas: Is there a key moment in your career that influenced your outlook on sustainable finance and the role of capital markets?

Waygood: There are many moments that have informed my approach. I see daily the good that money can do, and it is both fascinating and inspiring. As well as investing capital in companies that provide solutions to global problems, I've had firsthand involvement in using the legal and economic influence that sits within equity ownership to stop companies from manufacturing cluster munitions, extracting fossil fuels from World Heritage sites, or digging up indigenous people's tribal burial grounds in India to mine for minerals.

But one of the most important moments was in 2006 when Jonathon Porritt asked me to take a system view of the problem. Jonathon asked me to present to a course at the University of Cambridge Institute for Sustainability

Leadership [CISL]. His challenge to me was to try to cover the overall structure of the capital markets, highlight how they undermine sustainable development, and then point to changes in the system that would support sustainability. This was a hugely daunting challenge—particularly to cover all this in ninety minutes. But it turned into an absolutely fascinating exercise.

Since 2006, I have run this seminar annually at CISL, which has given me the opportunity to use the lessons learned from my PhD research and my practice within fund management while continuing to learn from the course's students. It inspired me to write Aviva's *A Roadmap for Sustainable Capital Markets*, which we presented to the United Nations in 2012. It subsequently fed into our Sustainable Capital Market Union advocacy at the European Union from around 2014. This directly led to the creation of the EU High-Level Expert Group on Sustainable Finance in 2015. The ensuing EU Sustainable Finance Action Plan has changed the structure of sustainable finance across Europe and is now influencing the rest of the world. I can trace elements of today's global policy framework on sustainable finance back to those hugely challenging questions Jonathan first asked me sixteen years ago.

Pagitsas: That is an excellent transition point to discuss the framework in which companies operate globally. What are the levers that influence how a company operates?

Waygood: The vast majority of the economies in the world could be described as capitalist. The economic growth model is focused on GDP [gross domestic product] growth, even though the architect of GDP, Simon Kuznets, said it shouldn't become the focus of policy at all, let alone the unique focus. GDP growth is driving every capitalist system in the world, possibly even others too. It is money that fills that growth.

With that money comes both opportunity as well as obligation. Particularly with equity investing in most Anglo forms of corporate law, you own the company, and you are the principal. As a result, they are your agent. They work for you. If you know what you own, and you can behave individually as a long-term owner, you can express support for good companies doing good long-term sustainable business practices. If you want, you can turn up at the annual general meeting [AGM] and participate in the parliament of the firm. That's an option and an opportunity. I'd argue it's an obligation because those of us who understand how difficult it is to run a listed business in very short-term markets know that the leaders of those businesses need vocal support from their investors, too. That's the kind of obligation investors have. The companies that are not listening at all are the ones that need greater scrutiny by their investors.

Pagitsas: How do large institutional shareholders like Aviva influence a company's business strategy, if at all?

Waygood: By the principles of the business, when company law was originally crafted, shareholders were put at the heart of company law, and they are the primary regulators of business. It was broadly assumed that shareholders would all know what they owned. In America, it is more common to know what you own in a pension or a 401(k). Outside of America, it is not common. Most of our eighteen million clients in the United Kingdom have no idea how their pension works, what it's invested in, or how their votes were used at the company AGM.

We do not live in a shareholder democracy, no matter how many times people tell you we do. We don't. The real shareholders, the ultimate beneficiaries, often don't even understand finance, let alone know they're shareholders or understand how to exercise democratic control in the parliament of the firm that they own. Most would have no idea that there is an AGM or a parliament at the firm. The problem is that this is not in the education curriculum in most countries.

There is a supply chain of capital—from you as an individual and through to your trustees, the investment consultants they use, the buy-side fund manager, the stock exchange, and then to the investment bank. Only when it reaches the investment bank does it start getting close to the company that puts your money to work to shape the future you wish to retire into, generating your return on capital employed and paying you dividends. Then, hopefully, through share price returns, the company may also be paying you interest rate on a loan you've given them, known as the *corporate bond*. Then it flows back to you, but as a shareholder, you are supposed to behave like an owner.

The fact is that the vast majority of people don't know how the fund manager used their vote on company issues. Just imagine what it would be like if our actual democratic process worked this way, with us outsourcing our vote to someone else without any real idea how they used it. It would lead to a deeply unjust economy run in the interests of the few.

Pagitsas: What is another challenge in the existing financial ecosystem?

Waygood: The system is short term. Ferociously myopic. As Gillian Tett, the editor of the *Financial Times Moral Money* newsletter, put it a little while ago, the system is focused far more on their next quarter than the next quarter-century. Because of that, we need to engage with the system, understand how it's structured and how it undermines sustainable growth. We then need to re-engineer and rethink it so that it supports sustainable growth. Otherwise, as my eleven-year-old daughter put it to me two months ago, "Money is destroying money. Is that right, daddy?" That is exactly what's happening. I wish I could have put it as clearly as that, as young as that.

Pagitsas: What will happen if we don't address these challenges?

Waygood: Our current economic growth model will probably destroy civilization as we know it. Not life on earth, not human, not humanity, but I believe the physical risks associated with climate change will start to erode civilization as we know it over the next few decades. The potential for extreme physical risks associated with climate change could destroy civilization as we know it. Money is destroying money. We need to harness money so that it supports long-term sustainable growth rather than destroying the growth opportunity for future generations. Sustainable development is supposed to mean "meeting the needs of the present without harming the ability of future generations." Of course, the current structure of markets is antithetical to that—diametrically opposed, in fact.

Pagitsas: What pressures are executives at listed companies feeling given this outlook?

Waygood: The leaders of listed businesses feel under constant pressure from market myopia and short-termism and feel incapable of investing as much as they would like to in their brand, training, morale, and long-term motivation because one of the easiest ways to improve your quarterly figures is by cutting the long-term investments that cost today but only deliver returns, often intangible, in the next eighteen to thirty-six months.

That paradigm is well understood and leads to a lot of people thinking long and hard before they list their company publicly. It is the case that markets do operate like that now. However, it doesn't support value generation over the long term. It isn't in the best interest of the company. It also arguably isn't consistent with the long-term fiduciary duty, properly understood. Yet, the largest owners operate like this because they, in turn, are under pressure to deliver results to their clients.

The fund managers who meet with the chief executives are under pressure from the pension schemes and other large asset owners to deliver good share price performance in the next quarter. It's even harder perhaps on their chief executives. It's not just delivering numbers that look good on the quarter. You then must have the rest of the market respond as well, and the companies that you've chosen outperform your peer group.

There are very few professions in the world where you can see how you're doing every second of every minute of every hour of every day against all your peers. Fund management is one of them. It breeds short-termism as it can be very, very, very hard to run money in such an environment. It can be extremely stressful.

Pagitsas: How does focusing on quarterly share prices, considered a core metric of financial performance, drive short-termism? Is there a connection to the role of stock exchanges in this conversation of sustainable finance?

Waygood: That frenetic pace of tracking performance and profits every second, every minute, every hour can lead to all sorts of unintended consequences. If Adam Smith, the economist who wrote *The Wealth of Nations* and coined the phrase "the invisible hand," could see the markets today, he'd be apoplectic to see how much damage has been done in his name, I'm sure.

Pagitsas: He didn't anticipate the speed of business today.

Waygood: No, he didn't. Stock exchanges used to be government institutions. They were quasi-regulatory. They maintained market quality. If you came to the exchange, you knew that the company always had a particular status, and it was governed in a certain way. Over the last fifty years, most of the largest listing venues and stock exchanges in the world have become for-profit. They listed themselves and made their executives focus on total shareholder return or earnings per share.

An exchange makes money from many different things, but it always makes it from exchanging stocks. High-frequency transactions are very much in favor if you're the executive of an exchange and looking to maximize your bonus next year. High-frequency transactions have absolutely shot up, but we've also seen a huge growth in short-term trading behavior. Much of this is algorithmically driven. They don't even understand what the company does, what the ticker stands for. They don't even need to know about the fundamentals of the company.

We've taken what was quasi-regulatory, and we've turned it into a machine for short-termism. Those of us that care about long-term innovation, research and development, and corporate performance over the next five, ten, and twenty years and understand how sustainability interweaves into them see the current structure of most business models is undermining the ability of companies to generate value, not even for future generations, but for current and future shareholders.

Algorithmic trading and high-frequency transactions all work against sustainability, yet most governments are allowing it to promulgate in the spirit of driving more liquidity onto markets. I'm not sure liquidity should be the god that it's held up to be. I think it's held up by those people who have a vested interest in making it generally assumed that liquidity is a good thing, so more is always better no matter the cost.

Pagitsas: Why do capital markets need to be engaged in discussing sustainable finance?

Waygood: The biggest question that the sustainability movement should be confronting is, how do you harness markets to deliver the UN Sustainable Development Goals [SDGs] and particularly the Paris Agreement? I say particularly the Paris Agreement because it's the one that, if not delivered on, will undermine the positive impact of all the other SDGs, and ultimately, could reverse economic growth and wipe out significant amounts of money. At an extreme, it may make money meaningless as it ceases to become useful as a store of value or a medium of exchange.

How you harness markets to deliver those important treaties is a question not yet properly addressed. Subsidiary questions of that are, what's the role of financial institutions? What's the role of companies in shining a light on that? Because, after all, the system was never designed to deal with these problems.

Pagitsas: In reviewing Aviva's current Responsible Investment Policy, it states, "We (via our asset managers) should actively exert influence over the companies we invest in to improve their sustainable environmental, social and government performance." Given the challenges you outlined, how is Aviva an active player and exerting influence to address ESG issues?

Waygood: It's important to address first that you don't change sectors and systems by only engaging with individual companies. You have to engage with those who structure the sector and the system and create the operating environment. That's my biggest, most important conclusion over the last twenty-five years in sustainable finance.

However, active engagement with individual companies is still incredibly important. We make no secret that we are very large, very long-term active owners of securities. We differentiate ourselves from passive investors, although we understand passive investment strategies as most of our clients will have them, too. We think actively exerting influence has several elements to it. First, being an active owner means understanding the fundamentals of the business and its strategy and actively seeking to understand what the company is trying to do to generate shareholder value. This means knowing its business model, understanding the caliber of the board and the quality of the individuals that are being elected, and knowing whether you wish to support them at the election or re-election at the AGM.

We take the vote as an asset itself, not just an obligation to vote. We pay a little bit more for voting share for a good reason. You get to exert influence over the company's future strategy. I want to do that positively and supportively, ensuring that those executives who are leading the company for long-term growth, for sustainable growth, can feel the support of their share.

Active ownership isn't just about engaging with the laggards and challenging them to be better. It's also about supporting the leaders. Active ownership and active influence recognize there's a partnership and that we represent the interest of the business principals, the beneficial owner. We are their agents. The board and their primary executives work for our clients in company law.

We know that we're considered the primary regulators of businesses, as all investors are. We take that job very seriously because the more that investors do that, the more the global markets work for all and not just a few. Sustainable long-term growth means looking after the employees, making sure customers, communities and the environment are not exploited, ensuring that the board adheres to good governance practice and is compliant with government codes, standards, regulation, and so on. Long-term ownership means raising concern if a company's business model is systematically exploiting any of its stakeholders to generate value.

When we're active, we support good practices just as much as we actively challenge poor practices. We will vote against the re-election of poor performing board directors, we will vote against excessive pay structures, and we will vote against under-performing management. We might escalate it to the chair. Obviously, different markets work in different ways, and the votes are not always available at every AGM. However, as a global owner and investor and by actively exerting influence, we bring together the discipline of active investment with the understanding of what it means to be an active and supportive owner.

Pagitsas: Is there a recent example that exemplifies Aviva's voting approach? What was the outcome?

Waygood: I've been involved with Aviva Investors' voting record over the last twenty years or so. Over that time, Exxon has had a shareholder resolution—at least one—almost every year. Those resolutions have been about climate change, human rights, employee labor standards, and the like. There were a series of critical resolutions in 2021 at the Exxon AGM. I'm very pleased with the outcome that those resolutions helped to make.

On climate change, Aviva Investors has voted in favor of these resolutions and against the Exxon board at almost every opportunity and are proud of the outcomes that we have played a small role in helping bring to fruition. Those outcomes include the addition of two new board members focused on speeding the transition to a low carbon economy by the oil and gas industry, required disclosure of Exxon's lobbying activities and spending, and accounting for if and how its lobbying aligns with the Paris Agreement.

It's important to bear in mind that getting to the point that we did, where the vote was as high as it was on the Exxon resolutions and where investors were winning new board seats and ejecting board members and the like, that's taken many years. We have taken voting action either against the chief executives being re-elected or against the pay structure for the best part of two decades. We have supported the vast majority of the shareholder resolutions. Now, that's unusual. It certainly won't be the record of any large US-based passive fund managers.

Fortunately, many fund managers now are recently realizing that they need to use their votes, as they are coming under much greater scrutiny to do so. Some of the very largest also realize that it is better if they do this publicly and openly. Their record is that their stance has shifted in a number of these areas. Our stance has always been that we need to make sure that each company's culture is sustainable and that the people employed throughout the business have got long-term growth in mind.

Pagitsas: Aviva doesn't just make equity investments in companies. It also invests in real assets like buildings, offshore wind turbines, and railroads. When Aviva is approached to invest in an oil pipeline, for example, what is the response? Is it a flat *no,* or is there some evaluation?

Waygood: It's definitely a thoughtful approach. It's analytical, as I'm sure you'd expect me to say. There are a couple of funds that take a pretty hardline view on certain activities. For religious reasons, for example, they won't ever own a particular investment. Apart from those funds, we will look at everything carefully and thoughtfully. We've got guidelines, but the exclusions are few and far between. However, we are not big friends of coal. That's an exception on the exclusion side.

One of the most important questions we ask is, "Is this going to be a stranded asset?" Either literally, because the physical piece of real estate or the surrounding area will be flooded or metaphorically because it's no longer being used and not generating the cash flows that it's supposed to cover its investment time horizon. Therefore, the DCF [discounted cash flow] is also wrong in the present value.

We also look at political risk. To what extent are the countries at either end of this pipeline likely to be stable and continue to generate the cash flows over the next fifteen to twenty years? That's a human rights question as much as it is an analytical one.

Related to how we assess climate risk, we've got people who work within the business who have a deep understanding of climate science. As an insurance business, we have a small army of actuaries who can give us insights into how historic assumptions might be changing. We also look out at the different sectors we invest in and have a sense of which ones are going to be more exposed to transition risk than others.

In particular, climate risk is asymmetric. You've got the risk of transitioning too fast and the risk of not transitioning fast enough. The former is called *transition risk,* and the latter is called *physical risk.* Climate risk is asymmetric because the physical risk is far, far, far greater. As I had mentioned earlier, the physical risk could be "civilizationally" challenging, but it also could lead to the demise of entire sectors, insurance among them. Over this asymmetric transition analysis, you've got time. If we don't change from an economic growth model that will result in the Earth heating by at least three degrees Celsius soon, the next few decades will seem relatively benign compared to the physical chaos of the last few decades of this century. If policymakers just focus on the period up to 2050, they will completely miss the much more significant challenges ahead. If not dealt with, physical risk will get exponentially worse over time.

Pagitsas: The multitude of time horizons is challenging, which sustainability leaders must grapple with since so many horizons intersect with each sustainability goal.

Waygood: Investment time horizons, valuation time horizons, and business cycles are different. Again, they're different to political cycles, which are, of course, different to geological time frames. When you're talking about climate change, it's not just geological anymore. It's happening around us today.

You're also talking about theoretical time frames. You're talking about decadal shifts, where we may have still yet maybe five to ten years to stop it. But you're overlaying that time on political risk, political time horizons, and business valuation horizons. In other words, how far out we assess and then place a value on the cash flows of a business. This is the tragedy of the horizon, I think, that our climate refers to. Each of those horizons is different. It's particularly tragic that we only have a short window to deal with the climate crisis.

The tragedy is that regulators, politicians, and the leaders of companies can't think over those decades or time frames to make this the most important thing they're focusing on today, yet it should be. That's the tragedy, but it isn't yet finished. The story isn't over. It might not be tragic. It doesn't have to be inevitable.

Pagitsas: I'm hoping that the story isn't finished for your kids and mine.

Waygood: Indeed. How do we turn this into some kind of epochal shifting, positive outcome? Well, financing has changed in the last, say, fifteen years. The United Nations created the PRI [Principles for Responsible Investment], and I was proud to be on the advisory group. Hundreds of people were involved in its creation. It was a group of people who, at the time, felt that finance could be harnessed to do good things. They were right, and it has been a successful framework to guide financing strategies.

ESG is absolutely massive today. This is significant because the people who created the PRI were right, and they've mobilized a lot of institutional demand for more patient ownership and longer-term voting. Many companies now get some ESG questions from some analysts because the PRI has changed the market. However, it's not enough. It's nowhere near enough to support chief executives who want to create sustainable growth over the very long term of the business cycle, and many business cycles, in fact. It's not enough, particularly when it comes to system change. And most importantly, it is not enough to correct market failures.

Pagitsas: Can financial institutions meaningfully engage and steward this system change?

Waygood: The word *stewardship* is used routinely, and every investor says that they'll be a good steward. However, very few can tell you what that means in the context of government bonds. Yet for large institutional investors, many of them including us, that's the biggest asset class where our money ends up. When we purchase government bonds, we are, on paper, owners of the country as shareholders. Clearly, we don't own the country, nor should we. You don't get a vote in the system. It's not a democratic tool.

There are democratic and diplomacy issues and accountability and transparency questions that arise when you engage, as a financial institution or as a company, within the political process. However, the engagement is legitimate for a range of reasons. Not least, because the policymakers themselves seek it because they tend not to understand the financial sector and the market in general. It's also legitimate because they are making the final policy decisions. It's also in our clients' long-term personal and financial interests for us to seek to promote a more sustainable outcome.

I find it odd that macro-stewardship, which encourages the correction of market failures via government policy, is often seen as potentially negative. I think this is due to questions of accountability and transparency. Through macro-stewardship, I believe we are trying to inform or participate in a debate around how governments can harness markets to deliver the SDGs and how they can internalize those externalities through new forms of fiscal intervention, market mechanisms like trading schemes for sulfur and carbon, and regulation and transparency measures for the end consumer. We can harness those mechanisms to ensure capitalism is innovative and promotes long-term growth.

These market corrections need to be seen through a lens of sustainable economic value generation. This isn't about free markets working unfettered, nor was it really ever. If you read [Milton] Friedman, he talks in his classic essay ["The Social Responsibility of Business Is to Increase Its Profits"] about board members adhering to generally accepted standards of business behavior. It is quite possible to see what he means in terms of ethics throughout.

If you want to do the right thing, but it pays more to do the wrong thing as a system, this is a market failure. This is defined as an unfettered market leading to a suboptimal outcome for society. Where this is the case, it means that those who structure the market—governments and their policymakers—need to restructure incentives so the profit motive is aligned with lots of sustainable growth. Free markets don't do that. Not at the scale that we need. You have one planet's worth of resources, but you are heading toward needing five to sustain the quality of life that we enjoy in Western Europe and in North America. We can't simply expect to go off and find four other planets.

We need to re-align markets so that they stay within the capacity of the existing ecosystem that we have. The diplomatic and technological challenge with that is somehow ensuring developing countries enjoy the same standard of life that we do.

Pagitsas: Do you feel that the UN Framework Convention on Climate Change [UNFCCC] Conference of the Parties [COP] is indeed the right vehicle to help bring in the structure that we need?

Waygood: Books could be written about that question. [laughs] However, the United Nations is the best forum we have. It gets criticized for being a talking shop, yet it was set up to be just that. Its consensus governance structure can be deeply dysfunctional. You have to have 196 member states align consensually on a document for it to go through at least in the Rio Process, the CBD [Convention on Biological Diversity], the Kyoto Framework, UN FCCC, and the Montreal Protocol. Because it's so consensus-driven, the governance structure for these treaties has ossified the possibility of a vast, prompt, scale response by the global community.

However, it is the best forum we have. While it's not set up to be particularly fast, it is set up to be inclusive and accountable. Perhaps this is the only pace we can transition at scale and at a pace that meets the many human rights and justice considerations that emerge. One of the things that we need to recognize from the perspective of markets is that foreign office officials, foreign ministers, and occasionally heads of state attend the UN General Assembly. It is not attended, typically, by finance ministers or central bank governors, although that has started to change. Mark Carney has impressively led this change. He has inspired more action from many central banks, the Financial Stability Board, the World Bank Group, and various other fora within the International Financial Architecture.

Pagitsas: How would you change the structure within which financial institutions play to better address climate change?

Waygood: 2021 is the first time the G7 and the G20 have both been hosted by the two co-hosts of COP. The closest was in 1998 when Germany hosted some of those meetings. We need to have a geopolitical alignment regularly, let's say every three to five years, at all the big ratchet meetings as they're called because the Paris Agreement has a ratchet effect within it. Ideally, the two co-hosts of COP make sure the G7 and the G20 mandate that organizations such as the International Organization Security Commissioners, the International Accounting Standards Board, the International Monetary Fund, the World Bank, and all the other multi-lateral organizations that are part of the International Financial Architecture are held to account for helping finance the Paris Agreement. The Paris Agreement is a political treaty, not a market reality. You don't run markets through political treaties.

Pagitsas: What would be the role of financial institutions in this ideal structure? What will happen if this structure isn't implemented?

Waygood: Every central bank and every finance ministry needs to change the incentives in their system, their own country, so that doing the right thing at the corporate level reduces the cost of capital rather than increases it. That means you need fiscal measures. You need market mechanisms. You need a whole series of market corrections because climate change is the world's biggest market failure. It will lead to the destruction of the market overall potentially. It could lead to civilizational collapse, as I've said.

We need to harness markets and their productive potential, including the research and development capacity they can drive. We have enough money out there, but the innovation that's now required isn't political, isn't a foreign office, isn't in the science. We've got a political agreement. We've got science which is now clear. The innovation we now need is in markets, and that means it's not the UNFCCC. It's the G7, the G20, and all the multi-lateral processes that drive the financial architecture. That's why I think the UN is the wrong place now to carry the conversation on climate finance to fruition. It does not have a powerful enough finance mandate.

It is not populated by people who understand markets or finance, specifically. There are an honorable few, I should be very clear about that, but it is not universal. To be fair, it's very unusual to find people in the policy sphere that understand markets. I contend that no one anywhere understands the whole of finance. No one individual. It's just too big. It's like trying to hold infinity in your mind. Many experts in finance understand a part of it with significant depth, but nobody understands the whole thing.

Pagitsas: If not the UN, then where? You said the G20, the G7. Will they take up the mantle, or are we waiting for some new entity to arise that will?

Waygood: Yes, and yes. While the G20 Green Finance Study Group was killed by the United States under Trump a few years ago, it's been reincarnated as the G20 Sustainable Finance Working Group. There's potential there, and it's for the G20. Now you may think, "Great, there's hope." It's a working group, not a study group, which means it can make policy recommendations. Wonderful. Then you look at it from the perspective that there are another forty other working groups that the G20 has in satellite around it. This means that the working group is at the beginning of a journey, not by any means at the end. Yes, the G20 will take it on. However, in the long-term, does it need to create new institutions? Yes. But I also think it needs to repurpose the whole International Financial Architecture.

Pagitsas: Is there an example of a body that has taken major restructuring work in the past?

Waygood: The OECD [Organisation for Economic Co-operation and Development], for example, was originally created to restructure Europe post the Second World War. Obviously, it's done that now. Should we not now think about the OECD's role in this new century's greatest challenge of all, which is climate change? It's the world's largest and best-resourced think tank, staffed by economists, many of whom understand climate change. Should we not be thinking about them owning Article 2.1c of the Paris Agreement and helping mobilize markets' money and finance advising? It is a club for rich countries, but that can be an opportunity for the developing countries, who need to know how to access it.

The OECD can be repurposed not because it's a good thing for money to fly out the gates of developed markets, but because we need to make sure that investment in climate solutions is made with the highest possible return on both money and carbon. And this necessarily involves increasing the investments into developing countries.

Pagitsas: Coming down from large institutions and to the individual level, how should one engage with CEOs and other non-sustainability management leaders to integrate sustainability into their businesses?

Waygood: One observation, I think, is that to win the hearts and minds of many CEOs, we need to make the business case. There's a short-term, medium-term, and long-term case. There's a short-term one around brand building. There's a medium-term one about having a healthy eco-efficient organization, which has the lowest cost, highest return. Then there's the long-term existential case—the one I make most here. Of the three cases, the long-term business case is very strategic and potentially problematic. Positioning your business on the side of the angels is challenging. Making sure you surf the tide of change is part of how businesses will create value.

Pagitsas: How can CSOs sustain themselves in these conversations about systemic change, engagement with regulators, and international frameworks while delivering on goals related to environmental and social issues within businesses?

Waygood: CSOs must maintain personal sustainability while maintaining an awareness and an anger about the sustainability issues without becoming numb to them because you read about the issues every day. You learn something catastrophically appalling most weeks, something new. Sustaining yourself over many decades in a career that seeks to address those issues without burning out, without feeling overwhelmed is the most critical sustainability challenge of all. We must sustain ourselves so that we can make sure that we do achieve critical mass.

Pagitsas: What advice would you give others to sustain themselves in this often dire march toward a positive future that sometimes seems to move further and further away?

Waygood: There are many books written about multiple aspects of this. One of the classic references is in *The 7 Habits of Highly Effective People* [by Stephen R. Covey (Simon and Schuster, 1989)], which talks about the need to continually sharpen the saw. As Covey says, it's not just the physical health challenge. It's the mental health one, too. It's the mental health challenges, I think, in sustainability that are toughest. When you empathize with future generations, and when you empathize with people in other countries and see what's happening to them, it causes existential anxiety. There's no question, it causes deep anxiety.

To not drown in sorrow, one has to connect with nature, go for long walks, take deep breaths, learn how to be mindful, and trust that with enough people thinking about this, we will solve it. We must hope that there are networks and forces beyond our own spheres; and that outside of our purview people are taking action, rethinking markets and re-engineering capitalism so that we find a path toward sustainable economic growth.

It needs to be faith, actually—faith in humanity and finding sources of hope. Personally, I have been incredibly lucky to participate in initiatives that have generated huge positive change in the financial system. So, yes, I do have faith in humanity and have certainly not given up hope.

Michelle Edkins

Managing Director, Investor Stewardship
BlackRock

Alexis Rosenblum

Chief Corporate Sustainability Officer
BlackRock

Founded in New York in 1988, BlackRock is a leading publicly traded investment management firm, providing a broad range of investment management and technology services to institutional and retail clients worldwide. As of year-end 2021, the company manages over $10 trillion in assets on behalf of investors worldwide.

Michelle Edkins is a managing director on BlackRock's Investment Stewardship team covering the Americas, Europe, the Middle East, Africa, and Asia-Pacific. In her role, Michelle is responsible globally for institutional relations, policy on investment stewardship issues, and communications related to BlackRock's investment stewardship perspectives and activities. She also serves on the firm's Global Operating and Government Relations Steering Committees.

© Chrissa Pagitsas 2022
C. Pagitsas, *Chief Sustainability Officers At Work*,
https://doi.org/10.1007/978-1-4842-7866-6_24

Prior to joining BlackRock in 2009, Michelle held investment stewardship roles at two firms in the United Kingdom. She was a managing director at Governance for Owners and corporate governance director for Hermes Investment Management. An economist by training, Michelle also worked in government roles in her native New Zealand.

An active participant in the public corporate governance debate, Michelle was named in the National Association of Corporate Directors (NACD) Directorship 100 "most influential directors and governance professionals" list for the past ten years. She is the chair of the International Institute for Sustainable Development (IISD), a former chair of the International Corporate Governance Network, and an Aspen Institute First Movers Fellow. She currently serves on several industry initiatives to enhance governance and sustainable business practices, including the US chapter of the 30% Club, a market initiative to increase the number of women on boards and in senior management, and the Sustainability Accounting Standards Board (SASB) Standards Investor Advisory Group (IAG).

Alexis Rosenblum, *CFA, is BlackRock's chief corporate sustainability officer. She is responsible for driving connectivity and accountability across the organization to ensure the firm succeeds on its sustainability agenda. Alexis partners closely with various departments, including Human Resources, Social Impact, Technology & Enterprise Services, BlackRock Sustainable Investing, and BlackRock Investment Stewardship, to develop and implement BlackRock's corporate sustainability framework, policies, and disclosures.*

Prior to her current role, Alexis was a senior member of BlackRock's Global Public Policy Group. She engaged with regulators and legislators to encourage policy outcomes related to financial market transparency and responsible growth of capital markets. Alexis was previously a relationship manager for BlackRock's Global Consultant Relations team, where she was responsible for developing and maintaining relationships with US institutional investment consultants.

Alexis is a CFA charterholder and is active with the CFA Institute. Alexis was named a Rising Star of Wall Street by Business Insider *in 2020.*

Chrissa Pagitsas: Alexis and Michelle, it would be helpful to start with an understanding of BlackRock. What is the scope of its services for clients and its over $10 trillion portfolio as of year-end 2021?

Alexis Rosenblum: BlackRock is an asset manager, and we manage money on behalf of other people. Those people are millions of pensioners and savers around the world, including people within pension funds that are managing money for teachers or individuals saving for their retirement, buying a house, or funding college.

Our services include investment management and financial technology solutions that are provided to institutional and individual investors worldwide. We invest our clients' money in every asset class—including cash, fixed income, equities, and alternative assets. We also offer a wide range of investment products and services to provide clients tailored investment and asset allocation solutions.

Pagitsas: Next, what is the scope of its sustainability strategy?

Rosenblum: When we're talking about sustainability and how it's integrated into what BlackRock does, it's based on our conviction that integrating sustainability into the investments that we make on behalf of our clients helps them build more resilient portfolios and achieve better long-term risk-adjusted returns. That's the foundational concept that our sustainability strategy is built upon. There are key activities within four pillars that we undertake to implement that strategy.

The pillars start with investment stewardship, which is what Michelle does. Once we invest in a company, we engage with its leadership to encourage the management of material environmental, social, and governance [ESG] issues and the disclosure of that information to us as shareholders. Michelle will touch more on this topic.

The second pillar is ESG integration, which is particularly important in the active investment management business, where we're not just tracking an index. Rather, we're trying to exceed the return of a benchmark. We believe that to be successful over the long term, our portfolio managers must integrate ESG considerations into their asset allocation decisions.

One of the metrics of success that we've been working toward for this pillar is the integration of ESG information into our active investment processes. This means that portfolio managers are actively considering ESG risk factors that are material when making investment decisions. A key support or enabler for the ESG integration pillar for us and our clients is our technology, where we're putting ESG data and building analytics into Aladdin, our proprietary risk analytics platform, to facilitate the understanding and quantification of ESG implications on investments.

The third pillar is our sustainable investment products, where there's an explicit ESG investment component or objective to those products. We are significantly increasing our number of sustainable investment products across the board.

The last pillar is how we manage our organization. We recognize that because we're an investor in many other companies on behalf of our clients, and we have views on how those companies should be managing material environmental, social, and governance issues, we have to do the same ourselves and lead by example. Those companies will look to us for a view on best practices. That's what I do, which is to make sure that through our disclosures and how we manage our organization, we're integrating the same best practices and leading by example in our operations and strategy.

Pagitsas: Michelle, let's dive deeper into the first pillar of BlackRock's sustainability strategy. What is its scope? How do you approach it?

Michelle Edkins: Within the stewardship pillar, we focus on governance first. A lot of people talk about ESG, but we believe that the governance component is the most significant, largely because, in our experience, companies with poor governance structures are very unlikely to appropriately integrate sustainability considerations into their long-term strategy, operations, and reporting.

It's important for companies to have that rigor around governance to achieve the impact of having sustainable business practices. Our focus is on who's in the boardroom and whether those directors have the right skills and experience to advise and oversee management in delivering the long-term strategy. Are they ensuring that the broad spectrum of risks and opportunities are being taken into consideration? Is ESG truly integrated into the overarching operations of the business, rather than being siloed to the corporate sustainability function?

Governance should enable a company to integrate sustainable business practices fully. When we have concerns about the approach a company is taking or the level of detail in its disclosures, that's when we would engage company leaders and potentially vote against the re-election of directors. As asset managers, we have a fiduciary duty to act in our clients' best economic interests. This is why we hold directors accountable for their role in representing the interests of shareholders and other stakeholders in the boardroom.

Pagitsas: It's an interesting point that you're raising, Michelle, about the role of governance. On the surface, governance may be seen as very simple. A firm's leader may say, "We have a board of directors and a chief compliance officer, so my company is well-governed," and check the box. What else is necessary to demonstrate good governance and stewardship to a firm's shareholders, clients, and stakeholders beyond those check marks?

Edkins: That's exactly it. It shouldn't be about the checkmarks so much as about demonstrating the effectiveness of each of those functions. It isn't just about having a board. It's about demonstrating that you have the right people in the boardroom and that they are actively engaged with senior management, understand the business and the dynamics of the sector, and so on.

Ideally, your board should be a competitive advantage, and what we're looking for companies to demonstrate is that they not only have all the checkbox components, but they can also give us confidence that all those functions are working as intended, whether it's audit and financial controls, the subcommittees of the board, or anticipating changing megatrends affecting their sector and adapting to them.

Pagitsas: My takeaway is that good governance comes from an actively engaged board that continually evaluates a company's performance against goals and how it is executing against those goals, operating within the current market, and preparing itself for future market changes.

Touching on the ESG integration into the active investment management business, the second pillar of BlackRock's sustainability strategy, what is an example of this integration in an investment management process or a fund?

Rosenblum: BlackRock manages more than 100 different investment strategies across a range of asset classes, so it's difficult to generalize, as ESG integration will differ by strategy. One example is where BlackRock invests directly in real estate and infrastructure. For these investments, ESG integration includes a detailed review of environmental and social factors, including those relating to environmental protection, pollution prevention, the conservation of local habitats and biodiversity, as well as the health and safety of employees and local communities.

Pagitsas: What sustainable investment products does BlackRock offer as part of its third pillar? How do these capital market solutions address the environmental, social, and governance challenges companies face and to which stakeholders ask for solutions?

Rosenblum: Over the past few years, BlackRock has significantly expanded the number of sustainable investment strategies available to clients. These include a wide range of investment strategies across indexing and active investment management. There are many different types of sustainable investment products, including those that incorporate ESG objectives broadly, to those that invest around a specific E, S, or G theme, or even products that seek to produce a measurable environmental or social benefit in addition to financial returns.

Perhaps just to cite a more intuitive example, one of BlackRock's active equity teams launched a circular economy investment strategy in partnership with the Ellen MacArthur Foundation in 2019. At its core, this strategy helps BlackRock's clients put their money to work in support of the transition to a circular economy and businesses that are evolving to a more sustainable way of operating. BlackRock also launched in 2021 two active sustainable exchange-traded funds [ETFs] focused on investing in companies that the investment team believes are better positioned to benefit from the transition to a low-carbon economy.

Maybe you are sensing that variety and choice is a theme here. That's by design as we recognize that our clients are diverse, they have a range of objectives, and frankly, there is a range of objectives and motivations that may cause them to seek out a sustainable investment product.

Pagitsas: Were these sustainable investment products initially met with skepticism by the market?

Rosenblum: Skepticism is probably the wrong word. But it's safe to say that sustainable investment strategies were more niche products for a long time. Today they are a central component of many of our clients' investment portfolios. Just to give you a sense of the magnitude of this shift in a relatively short period, in 2018, BlackRock managed $52 billion in sustainable investment strategies. As of December 31, 2021, that figure is up to $509 billion. Over the course of 2021, BlackRock saw a record $104 billion of net inflows into sustainable investment strategies. This shift demonstrates the scale of the change that is underway in investor behavior. The thing that probably surprised people the most is that the momentum toward sustainable investing accelerated right through the COVID-19 pandemic. Previously, the conventional wisdom was that in good times investors would focus on ESG but that in bad times ESG would be deprioritized. That presupposition was clearly disproven by the pandemic.

Pagitsas: The fourth pillar of the strategy is about BlackRock's own operations. What is the scope of these activities? What is an example of a new operation resulting from the commitment to sustainability?

Rosenblum: This pillar is all about making sure we lead by example in how we conduct ourselves as an organization. That runs the gamut from the governance and incentive structures we put in place and how they align with our long-term strategy and objectives to how we manage the environmental footprint of our physical operations and how we show up and support our employees and the communities in which we operate. Our commitment to transparency is critical and fundamental to this work, particularly since, as investors on behalf of our clients, we need more and better-quality data from the companies we invest in to help us effectively integrate and quantify material ESG considerations in our portfolios. In that sense, we have an extra incentive to produce good corporate sustainability disclosures, given the commercial imperative to set a good example here.

We are also thinking critically about who we choose to do business with as an organization by building out our vendor sustainability and diversity program. We're thinking about how we can encourage the same types of behaviors and management of material risks and opportunities that we do in our own house. The vendor program will encourage our business partners to measure their progress around diversity, equity, inclusion, and environmental sustainability and put in place goals and strategies on key challenges or areas for improvement. In this regard, we've moved away from simply setting minimum expectations for who we will do business with, but also encouraging the adoption of best practices with those companies we do choose to do business.

Pagitsas: These four pillars are ultimately overseen by BlackRock's board. How does the BlackRock board engage on the firm's sustainability strategy, outcomes, and goals?

Rosenblum: Our board oversees our firm's entire long-term strategy, of which sustainability has become a key pillar. Within the realm of that oversight, the board has overseen the evolution of the sustainability strategy from an important focus to a critical and central component of our business strategy which has accelerated in the past two years.

In relation to governance enhancements, we have made oversight over the sustainability strategy much more explicit than it had been in the past. In January of 2020, we explicitly incorporated new sustainability oversight responsibilities into the charter of a board committee, the Nominating, Governance, and Sustainability Committee. The committee was also renamed to incorporate its oversight of this area into its name. The committee's oversight scope includes investment stewardship, which is one of the most critical components of that strategy, my work which is the corporate sustainability work, our corporate disclosures around ESG issues such as our SASB and Task Force on Climate-related Financial Disclosures [TCFD] reporting, Social Impact, which is our philanthropic work, and oversight for public policy activities and trade associations.

The board's role has been an evolution over time. When you think about the board's engagement in terms of the elevation into the strategy, the acceleration of our work, and the delivery on many of our commitments that we have set out, the outcomes demonstrate the strong role that oversight and governance can play.

Edkins: The importance of ESG at the board level is reflected in the two most recent additions to the BlackRock board, Beth Ford and Kristin Peck. They were announced in September 2021. Both bring direct hands-on sustainability experience from their current corporate roles.

Pagitsas: Is ESG or sustainability a core competency that you expect to see in every director in the companies that BlackRock invests in or only in some directors?

Edkins: It's interesting in the sense that in the United States, there is still a debate in board circles about whether sustainability even does rest in the boardroom or whether it's solely a management issue. Our view is if it's a significant or a material driver of value or risk, then it rests in the boardroom. We have communicated that view for a few years now. We sense that there is a growing appreciation of that expectation from investors because it isn't just BlackRock that thinks that.

It is important that the board and management define the key sustainability-related drivers of value and risk because the drivers depend on a company's business model. For some companies, water might be a significant driver. For others, it might be energy efficiency where they're really going to move the needle. For others, it will be supply chain and human rights protections. For most companies, it will be a combination of many things. It is homing in on the most significant drivers of value and risk and ensuring that within the board, and through the combined skills that the group of directors brings, there is the requisite expertise on those key risks.

It isn't to say there must be a climate specialist, a human rights specialist, and a cybersecurity specialist on the board. We would expect the directors collectively—because of the other roles they have had—to have a level of fluency in those types of issues that enable them to ask management the right questions and understand what's put in front of them. Given the complexity of most business models, it's problematic to ask for a single director who brings that single focus on each of these key issues.

As I heard someone else describe it, which I think is a nice description, every director will have a major and a minor. The major for all directors should be business, financial, and operational expertise. Then, each director's minor should be complementary and clearly aligned with the company's strategy, that is, the long-term future strategy, not the historical strategy, and some of those key drivers of value and risk. Not all of them are going to be sustainability-related either. To be clear, some of them could be financial or market related.

When it comes to demonstrating clear fluency or operational experience in sustainability matters, our observation is that there is a generational shift happening in boardrooms, and more of the newer directors, including first-time directors, have had more hands-on experience in their operational executive roles on sustainability-related matters, as those matters have risen up the corporate agenda.

Pagitsas: How does BlackRock apply ESG principles to its investments through company engagement?

Edkins: Incrementally. It has been a journey. We've had a stewardship function at BlackRock and its preceding companies for well over two decades. We've been very focused on governance for a long time. The transition that we've seen over that time is a broadening of focus from pure proxy voting to much more interaction with companies, meaning direct engagement. The breadth of the issues discussed in those engagements has also increased year on year.

Generally, our approach has been to include emerging issues in our engagement conversations as they become more commonly understood as business risks or opportunities. For instance, on climate risk, in our engagement conversations, we ask companies how they're thinking about climate risk and what they anticipate will be the impact on their business of regulatory or physical risk and so on. Based on those conversations and our own research, we move on to reflecting concerns in voting. In the common vernacular, we socialize the issues in our published stewardship policies and our engagements. Then we reflect them in our voting analysis and decisions, which may initially result in increased votes against management.

To use climate as a tangible example, we began engaging on climate risk around 2015. We would ask how a company's operations were exposed to climate risk. That was a tangible basis for a conversation about risk and how the company intended to mitigate that risk. Then we might ask for more disclosure about how the company is anticipating and responding to climate risk.

In our early engagements, the response from companies was, "What do you want us to disclose?" We would say, "Well, you know the risks in your business. We believe it's important for you to disclose what you think shareholders need to understand so that they can determine that this is a well-managed risk." Then, thankfully, the TCFD was formed, for which BlackRock was a founding member, and provided a framework for such reporting. Now, we're saying, "Please report to the recommendations of the TCFD." These steps are incremental because you have to bring people along when you ask for change.

Pagitsas: What has driven the increased focus by investors on sustainability issues, including climate risk?

Edkins: What has changed markedly in the last decade is investors' understanding of the material sustainability-related drivers of risk and value and how they may impact performance over time. To be clear, this is not investors bringing sustainability issues to the attention of companies. Companies have been dealing with these issues for a long time because if they are truly material, they will be key to their ability to operate. Therefore, they will be built into their management systems. Companies have, over time, increased their disclosure on sustainability issues, which has helped investors' understanding.

To use The Coca-Cola Company as an example, water is an essential input for them. The company's management must demonstrate that they use water efficiently, especially in water-stressed places, which is a lot of the United States. If they can't, they will not get a license to operate. Demonstrating efficient use of water, a key input, is not just an environmental issue. It's an existential one. Other ESG issues can be quite opaque, but Coca-Cola and water is a simple example and is easy to connect, I think.

Pagitsas: It's a great point because some businesses have material issues but are not as tangible as a water issue. An executive may say, "Climate change is too broad an issue, so it's not material to my business." Would you respond to them by explaining the business case for addressing climate change?

Edkins: Well, it's not for us to make the business case to them. It's for them to demonstrate they are managing their business in the right way to protect the long-term interests of our clients as shareholders.

Pagitsas: Does the shorter-term, quarterly nature of financial targets hinder adopting and achieving longer-term, decades-long environmental and social goals?

Edkins: Balancing the achievement of long-term business outcomes and near-term financial targets have long been a focus for corporate leaders. There isn't really a distinction between balancing making investments in a new line of business that may take several years to deliver returns and this year's results or between ensuring robust and sustainable operating practices that minimize the risk of product or processing failures. Those trade-offs are part of corporate management. As long as a company's leadership sets out clearly what it is doing and why it is in the best interest of the company and its shareholders, it should get support from shareholders and other stakeholders aligned with that approach.

Pagitsas: How has the importance and scope of ESG considerations and principles and their integration into BlackRock businesses evolved since sustainability was first brought into the business strategy?

Rosenblum: This conversation is making me flashback to one of the first conversations I had with a mentor of mine, who was the one who hired me to start a corporate sustainability function for BlackRock. One of the things he said to me was, "Sustainability can become the whole company very quickly. At the same time, environmental and social issues can be defined very narrowly, so you're going to have to figure out how to draw the boundaries around it."

As I set up the program and the framework for putting the lens on BlackRock and thinking about ESG, a couple of things became apparent. First off, in many ways, ESG is a series of topics that felt like no one knew what to do with them, so we threw them all together. However, they're very different things. The internal stakeholders are very different. For example, the people working on environmental topics, such as our facilities team and our technology team, are very different from the people who are working on social issues like diversity, equity, and inclusion or our philanthropic work, and who in turn are very different sets of people from those who are doing our sustainable investing work. That's something that has become apparent through this journey. We tend to put all these things together, but they're very different sets of issues and skillsets.

Pagitsas: Managing the energy strategy of your own office space is a very different exercise than engaging with a company's executive leadership about the energy strategy of their manufacturing plants.

Rosenblum: Yes, exactly. The other thing that we had to grapple with was how we define corporate sustainability for BlackRock. How do we get specific about what we're talking about and managing when we say "ESG"? For BlackRock, we ended up looking at our businesses, processes, and material issues. We ran a very traditional stakeholder assessment, where we tried to understand the views of our clients, employees, shareholders, and this vast other category that we call "society." Some people call this a "materiality" assessment. Coming from a financial services background, I have a hard time with that terminology since the concept of what's material or not can mean different things in different contexts. Hence, I prefer to call it a "stakeholder" assessment.

We worked with the key sets of external stakeholders to understand their views. It was a super interesting process through which to take in external feedback. One of the things we had to grapple with as an asset manager is that our ESG issues are often the issues of the companies we invest in on behalf of our clients. Our stakeholders want to know about other companies' ESG issues because, as investors, we are associated with the companies in which we invest.

Throughout the process, I kept looking at Global Reporting Initiative [GRI] graphs that companies have generated. It graphically shows which environmental, social, and governance issues are important to a company. I kept saying, "Okay, what's the formula? What's the function for mapping these graphs?" We did a lot of outreach to organizations and others who had experience with stakeholder and issue mapping. I found that there's not a lot of science beyond intaking the feedback from your stakeholders.

You then have to make your own assessment of where those issues fall in the hierarchy of priorities to address and integrate into business processes, policies, and reporting. This is an important area where more work needs to be done generally. We also looked to key external standards for guidance, including the SASB, TCFD, and the UN Global Compact principles and Sustainable Development Goals [SDGs].

Subsequently, BlackRock launched the Center for Stakeholder Capitalism to do research around stakeholder capitalism. There's an opportunity to think about how we get a bit more scientific in how we rank the relative importance of different ESG issues to companies as a whole segment of the professional world that focuses on sustainability. For BlackRock, we were able to come up with a list of key ESG topics that we disclosed and used to build our whole framework of what ESG issues and data we now report on.

Pagitsas: What is the final list of ESG issues for BlackRock identified by the stakeholders?

Rosenblum: As noted in our 2020 sustainability disclosures, they are as follows—business ethics and conduct; sustainable investing and stewardship; climate-related risks and opportunities; employee diversity, equity, and inclusion; employee health, safety, and well-being; board composition; public policy and political activities; human rights; natural capital-related risks and opportunities; selling practices and product labeling; supply chain management; and community relations and social impact.

Pagitsas: What was an area of critique or development for BlackRock from its stakeholders?

Rosenblum: One of our biggest challenges is that our stakeholders are diverse. They have a range of opinions, and sometimes those opinions are diametrically opposed to one another. I'd say, overall, a key theme was the need to be more transparent about what we do and why we are doing it that way. That's one reason why Michelle's team has increased their transparency and reporting around our voting and engagement activities so significantly over the past two years.

Edkins: Building on what Alexis said about identifying what to focus on as a company, particularly if you're just starting in pulling your sustainability narrative and data together, we strongly recommend that companies look at the resources published by SASB because they've done a lot of the work for each industry to map material sustainability risks. They have set out what on their assessment are the key, material, or business-relevant sustainability factors and suggested metrics for reporting. From an investor perspective, focusing on a relatively limited set of data and metrics on a clearly sector-specific basis is hugely helpful for peer comparison and our ability to understand the dynamics of a company's business model.

Pagitsas: Given these material issues, what top-level, public goals or commitments has BlackRock tied to them?

Rosenblum: BlackRock has made a range of commitments across ESG issues broadly as well as in our sustainable investing and investment stewardship efforts. A few top-line ones are integrating ESG factors into our active investment process, increasing the number of sustainable ETFs we offer, enhancing the transparency of our investment stewardship activities, increasing engagement with the companies in which we invest on climate-related risks and opportunities, and setting science-aligned emissions reduction targets for our corporate operations.

Edkins: Additionally, human capital is an important contributor to BlackRock's long-term success, and the leadership from the board down has said we need to have a diverse and inclusive workplace. Last year, in light of the Black Lives Matter movement and a real spotlight on social justice and equality and equity, the firm publicly said, "We haven't made as much progress as we should have, and we're going to double down on this."

Therefore, we are focusing on increasing the representation of Black and Latinx employees by thirty percent in the United States and doubling the number of Black and Latinx senior leaders in the United States. We've achieved many of our goals and have made considerable progress toward others, but suffice it to say, we still have a lot of work ahead of us.

Pagitsas: Alexis, to paraphrase your mentor's comment, "Sustainability can be very broad, and it can also be very narrow." It raises the question of who should lead on sustainability at a company. Do you anticipate that the head of the sustainability role will exist in ten years, or because everybody understands and integrates sustainability into their daily business decisions, we won't need one person to lead sustainability?

Rosenblum: What I would say is for companies that have been leading the way on sustainability, it's not just one or two groups who are doing sustainability. It's about ensuring the concept of sustainability proliferates across the organization in terms of expectations and expertise that's coming into the organization. Even for us, it's diffusing internally very quickly.

There will still be room for specific sustainability professionals who monitor, report, and ensure the organization maintains best practices from a sustainability perspective. It might just not be that the center of power for sustainability is necessarily concentrated in one place. Companies that are just starting on that journey, and certainly, there are plenty of them that are in that boat, they may be five to ten years behind in terms of how it proliferates across the organization, and that's perfectly fine. We found that we needed a concentration of sustainability professionals to coherently articulate what we're doing from a sustainability perspective. That's certainly a fine approach, but I think for a lot of companies, over time, it will just stop being a separate and distinct function.

Pagitsas: A related organization question is, where does the corporate sustainability team sit within BlackRock?

Rosenblum: Where the sustainability function sits is a big question mark for many companies. My team has moved around. We started in the finance department, and now we're in a new department called External Affairs. We make sense in a lot of places because what we do is so integrated with the rest of the company.

When Blackrock's External Affairs group was launched in June 2021, my team, corporate sustainability, joined the External Affairs function alongside BlackRock's social impact team, public policy, and BlackRock's new Center for Stakeholder Capitalism. External Affairs comes together because we're all engaging with our stakeholders—whether it be policy makers, the communities in which we operate, employees, and many more—and trying to understand their views and integrate that feedback into our work across the organization.

When I was in the finance department, it was all about disclosure and controls. When we think about where the SEC [US Securities and Exchange Commission] will go in terms of regulation, having controls around our sustainability data and disclosures is extremely important. It makes a ton of sense to sit within the finance department from that perspective. And while my team no longer sits within finance, we still work closely with the finance team, as controls and process continue to be an important focus of our corporate sustainability disclosure work.

Pagitsas: Michelle, over your tenure at BlackRock, has there been a shift in the approach to investment stewardship, and what has driven the shift?

Edkins: Over the last decade or so, we've increasingly built sustainability considerations into our engagement and, more recently, voting. In 2020, the firm went out with a strong message in recognition of sustainability risk—and climate risk in particular—as investment risk. Therefore, as a core part of our investment approach going forward, it gave impetus to what stewardship was doing in this area. As a result, we built more specific policies, particularly in relation to climate and the energy transition. We will continue to integrate those factors in our engagements and our voting.

Enhanced disclosures are key to the whole process of building more sustainability factors into what we do. Companies are reporting more, but there's also more reporting *about* companies with information that the companies themselves don't issue. That's an important component that a lot of companies don't necessarily appreciate. Investors, especially investors like BlackRock, who have teams that do quantitative analysis, can pull a lot of data from a range of sources that tell a company's story. A company's report is but one input into the analysis.

The stewardship team benefits from that use of technology by teams within BlackRock and external research, and it helps broaden our engagement dialogue. On top of that, there's just a much greater appreciation since the financial crisis of the impact of companies and investors in the global economy and the importance of demonstrating you're taking a thoughtful approach to your responsibilities.

Pagitsas: Continuing our discussion on BlackRock's internal sustainability approach, where there is internal resistance at BlackRock to a new ESG-related activity, what does it take to overcome it?

Edkins: It all comes back to having commitment and leadership from the top levels of the firm. We can't overstate how important it is to have CEO leadership and a clear and readily understood strategy on key sustainability matters at a company to get it truly embedded, and even then, you may never get it fully, fully embedded. You may never get 100 percent buy-in. But, as it becomes increasingly part of the fabric of the organization, it becomes increasingly difficult not to be part of the effort.

Rosenblum: That's right. We've been very fortunate because our CEO, Larry Fink, has been so upfront and willing to take a leadership position on some of these topics, especially through his annual letters to CEOs. As other companies think about investing in sustainability strategies and practices, the buy-in and the voice from the senior executive leadership, including the board, is super important. It definitely has influenced the way that our strategy in ESG issues has come together.

Pagitsas: Yes, it does takes courage and conviction from the board and executive leadership to shift to an ESG-integrated future. ESG must be signaled from the top, across the company, and consistently. Thinking now globally, are companies in Asia, Latin America, Europe, and North America thinking about sustainability the same way?

Edkins: I think it's difficult to generalize. The drivers in each region, market, or culture can be different. We were having a conversation internally today about certain markets in Asia. In Japan, this concept of companies having social responsibility is very long held. In Japan, companies have been very focused on delivering on the UN Sustainable Development Goals relevant to their business models in a way that you are unlikely to hear from an American company.

There's a lot of variation driven by culture, corporate legal constructs, and market norms that affect what companies prioritize or emphasize, the gaps between stakeholder expectations and corporate responsibility, and so on. This is why, for stewardship, we are aiming to have high-level globally consistent principles and areas of focus that we apply through a local lens. For example, looking at diversity and diverse leadership is hugely important, but that's a very different thing in North America than in Latin America or Asia.

Pagitsas: Similarly, there is a wide spectrum of corporate governance standards globally. How do you and BlackRock leadership engage on governance and policy across this spectrum? What are the challenges?

Edkins: BlackRock, on behalf of our clients, is invested globally. Therefore, across multiple aspects of our business, we believe that we have a responsibility to try to ensure that the market environment within which we are investing is aligned with our clients' long-term economic interests. In the context of our investment stewardship activities, this primarily means clearly defined shareholder rights consistent with corporate transparency and accountability. The quality of corporate disclosures is important because that's the basis on which investors make decisions, as is the encouragement of stewardship or active oversight of companies by their investors. I'm particularly talking about shareholder rights because of my role, but others within the business would talk about property rights in relation to real assets and so on.

We think it's important to have a voice in the policy debate because we believe it's important for ourselves, and for practice generally, to continually improve. Globally a lot is being done to advance sustainable business practices, but the investor's voice is often left out. The listed company has quite a loud voice, public policymakers have quite a loud voice, corporate advisors have quite a loud voice, and civil society groups tend to be part of that mix. Yet, the investor's voice is often brought in at the very last minute or not heard at all.

Our engagement starts with our top leadership and continues through multiple leaders and channels. It includes senior leaders attending global gatherings like the UN Framework Convention on Climate Change [UNFCCC] Conference of the Parties [COP], the World Economic Forum in Davos,

Switzerland, and the International Monetary Fund [IMF] meetings. Beyond that, as practitioners, we participate in focus groups and projects. For instance, we were part of a working group led by the OECD [Organisation for Economic Co-operation and Development] on responsible business conduct, and we responded to the IFRS [International Financial Reporting Standards] consultation on whether it should set up the International Sustainability Standards Board, to name a couple.

Regarding channels, one is informal dialogue and relationship network building, and the other is responding to formal policy consultations. In the consultation cases, as a firm, we will determine which colleagues are best placed to contribute to developing our response. To give you an example of this, a few years ago, the OECD made clear that its global guidelines for multinational enterprises on responsible business conduct applied to investors. Yet investors hadn't been part of the original process to develop the guidelines, so they don't relate directly to how investment decisions are made. In 2016, the OECD launched a working group of which BlackRock was a member along with other investors, including banks and insurers, to translate the guidelines for multinational enterprises into a more actionable set of guidelines for investors that relate to the corporate guidelines for responsible business conduct.

Pagitsas: Is there any pushback, Michelle, on BlackRock being vocal about the sustainability or ESG strategy, and where does that come from?

Edkins: This is the delicate balance that we try to achieve, which is speaking out to advance our clients' interests without leaving a vacuum into which people can say, "BlackRock is trying to be quasi-regulatory" or similar. We seek to contribute thoughtfully to the debate without people perceiving that we're trying to dominate it. This is where we must roll our sleeves up and contribute at the working group level, so we are part of a shared conversation.

Pagitsas: It seems there must be self-awareness about a company's role in the international regulatory, standards, and policy conversations, and you must calibrate your engagement to who you are, how you're perceived, and so on. It's a thoughtful engagement.

Edkins: You make an important point. We are disciplined about participating in initiatives or consultations where we believe we have something to add. We don't just speak up for the sake of speaking up, but when we do, it is because we have a perspective that we think isn't in the mix. Whether it's as a longtime shareholder on behalf of our clients, whether it's as a real assets investor, whether it's significant expertise in renewable energy, whatever it is, we're focused on where we can bring something that might help move things forward in the interests of our clients.

Pagitsas: Where do you anticipate BlackRock will continue to engage for the long-term because the investor perspective is absent?

Edkins: I'll speak to the section of the BlackRock activities that I'm very attuned to, investment stewardship. It's quite multifaceted in my experience, particularly given our position as a globally diversified investor across multiple asset classes. One area of focus is to advance practices on sustainability reporting and help companies focus and deliver information that investors need with a relatively low reporting burden on them.

There's also a fair bit to be done in terms of changing the narrative that companies are only focused on shareholder returns. There's been this false dichotomy for the twenty years I've been in this role. It's not about shareholder returns at the expense of everything else. As I'm sure is going to come out loud and clear in your book, without advancing and protecting the interests of the key stakeholders that a company depends on for its success, it doesn't— indeed can't—serve long-term shareholder interests.

It will take long-term engagement to move beyond this false dichotomy. In advancing the interests of their key stakeholders, companies can do things that are in their control. Still, they will need predictable long-term public policy to take what are often multi-year decisions, or decisions that have multi-year impacts and investments associated with them, to help in the areas where they can control outcomes that address the interests of their key stakeholders, including their shareholders.

Pagitsas: The largest opportunity to engage with regulators and the private sector on climate change has just concluded at UN COP26 in Glasgow. What major commitments did BlackRock advocate?

Rosenblum: For BlackRock, COP26 was a moment to strengthen our engagement with policymakers and officials as well as meet with clients and opinion leaders. We went to COP to advance discussions around TCFD adoption worldwide, build on the COP themes to showcase actions, and help shape policy outcomes around the transition to a net-zero economy. Ultimately, a successful and orderly transition is better for everyone, including BlackRock's clients, who are planning for long-term goals.

Pagitsas: In addition to investor stewardship, sustainable financial products, and engaging with policymakers, BlackRock has other tools in its toolbox to address environmental, social, and governance issues. Alexis, what is the scope of BlackRock's philanthropy? How is philanthropy used to advance sustainable issues?

Rosenblum: Yes, philanthropy has been around for a long time at BlackRock. We have been doing it strategically since 2106 through our Social Impact program, where the focus is on how we can have an impact on the communities in which we operate. The work has two pillars—our foundation and our employee engagement program. In February 2020, we took a big step forward in our philanthropic work when BlackRock donated close to $600 million to fund the firm's long-term social impact efforts and establish The BlackRock

Foundation. The foundation helps people beyond BlackRock's core business build financial security and participate in the transition to a low-carbon future. Coupled with our employee-directed philanthropic efforts, our philanthropy seeks to advance equitable and sustainable economies and communities. This is an increasingly important focus for the firm and our leaders in terms of how we show up for the world around us.

On the climate transition front, in particular, we view philanthropic capital as an underutilized but very important tool to attract the investment needed to achieve the scale of capital necessary to fund the global energy transition and the global transition to net zero. That's because there are so many new technologies that need to be developed and scaled to achieve the transitions. Many are at an early stage or still not market viable that they may not attract the type of institutional capital that BlackRock's clients would typically want to put toward investments.

One example is the BlackRock Foundation's recent commitment to give $100 million, which is the largest grant to date, to the Breakthrough Energy Catalyst program. It is a new philanthropic venture whose aim is to cut what's called the "green premium." At times, it's more expensive to buy the more sustainable product. An instance of the green premium is sustainable aviation fuel, which is one of the technologies that the Catalyst program is working towards accelerating market adoption by closing the cost differential between sustainable aviation fuel and traditional jet fuel while also making it more widely available.

The program is looking to invest in technologies that will help to cut that green premium over time as fast as possible. It's currently focusing on four technologies, which are direct air capture, green hydrogen, long-duration energy storage, and sustainable aviation fuel. It's focusing on those areas initially because of how important those technologies are to the transition.

Ultimately, our view is that philanthropic capital, whether ours or from other funders, plays almost in the first loss position, where that capital is willing to take a much greater risk because it's not looking for a financial return as a primary objective. It's looking for a social or environmental goal as its primary objective.

Pagitsas: What would you advise a CSO if they are evaluating launching a philanthropic solution?

Rosenblum: I would start by saying that I think that the first job of a CSO is to make sure that the company itself is operating sustainably and is managing material, environmental, social, and governance risks. While philanthropic capital is certainly important and wonderful for whatever causes one might pursue, if you don't get your own house in order, it's not going to be a viable solution for the company that's not operating in a sustainable manner. That's where companies have to begin.

Pagitsas: Michelle, we've talked about the role of the board of directors and senior leadership, and now the role of philanthropy. Broadly speaking, what is the role of private companies in the transformation of capital markets to a more sustainable future?

Edkins: Something BlackRock states explicitly, and I believe strongly, is that the private sector has the power to effect change. We see some innovative things being done by companies that we aren't really in a position to acknowledge other than through voting in support of management. The nature of our stewardship work means we tend to focus on the companies that are a little lagging, a little behind the ball, to encourage them to move forward.

I would love to get a mechanism beyond BlackRock that acknowledges and rewards the companies that are innovating and taking the risks that will change the bigger picture. Arguably, that's what capital markets will do. It's also nice to acknowledge, just by recognition, that innovation happens within public companies and private companies, along with risk-taking, bringing people along, and creating a purpose-driven community. All those things, I think, are critical to the significant change that we're going to see in the next couple of decades, and I am very optimistic about the role of public companies in that.

Pagitsas: Alexis and Michelle, let's shift our conversation to your careers and philosophies as executive leaders. Alexis, what was your professional journey to becoming chief corporate sustainability officer at BlackRock?

Rosenblum: My first role was in BlackRock's institutional client business, working with investment consultants who serve pension funds and insurance companies. I learned the business working with our clients and covering many different products. I joined BlackRock in 2010, which was right when the Dodd-Frank [Wall Street Reform and Consumer Protection] Act passed, and regulators were focusing on addressing the financial system's role in the 2008 financial crisis.

As my academic background is in public policy, I always wanted to bring that back into my professional work. So, in 2012 I joined our public policy group and represented BlackRock to regulators and legislators in Washington, DC, and around the world. Toward the latter part of my seven-year tenure on the public policy team, those discussions morphed into a dialogue about corporate governance and ESG issues. I spent a lot of time working with Michelle's team to understand their perspective from a financial materiality perspective on the management of material ESG issues and how a company is performing relative to their goals, TCFD, SASB, and other standards.

Going back to our earlier discussion, ESG includes very disparate topics, which was made apparent as it became increasingly challenging for us as a company to answer the question, "What is BlackRock doing from a corporate sustainability perspective?" Unless you had twenty people on the phone to try to explain, you couldn't do it coherently. I raised that concern with our senior leadership and that we had an opportunity to accelerate how we were addressing ESG issues as a company so that we could start to play a leadership role, lead by example, and show the companies with whom we were engaging on the stewardship front what good looks like from an investor perspective.

In 2019, Barbara Novick, a vice chairman and co-founder and a key mentor and sponsor throughout my career, and our CFO, Gary Shedlin, encouraged me and the head of our investor relations team, Sam Tortora, to establish a corporate sustainability function for BlackRock. Sam had been facing the same sets of questions and challenges I had faced with policymakers but from shareholders. After that, my role was created. Shortly thereafter, we as a firm significantly accelerated our sustainability strategy across our investment capabilities, which further validated the reason and the need for the company to have a function that's specifically dedicated to working this out through the company's own organization.

And now, as regulators increasingly focus on this area and decide on disclosures and what companies need to be doing, it's full circle for me. I find myself in more and more policy discussions. But I'm coming at it from a different perspective as a corporate issuer, having done this disclosure and thought through these issues for ourselves.

Pagitsas: Michelle, what was your professional journey to becoming managing director of the Investment Stewardship team at BlackRock?

Edkins: I trained as an economist and went from university in New Zealand to the New Zealand Reserve Bank, which is the country's central bank. The role was very, very econometrics oriented, and I thought, "Well, this is fun." While it was a fun degree and experience at the bank, I wasn't sure it was what I wanted to do for the rest of my life. Basically, it was just too much math. So, I said, "Let's find something else." I left there to work for the British High Commission, which is the diplomatic service in a Commonwealth country between Commonwealth countries.

I worked as a policy officer there, mostly explaining the impact of New Zealand's significant deregulation efforts on the economy. That role was too much the other way, deeply steeped in politics, policy, and so on. I left there to return to the United Kingdom, where I'd worked as an au pair between school and university. I saw a corporate governance officer job advertised in the *Financial Times*. I had no idea what it was. About five years earlier, the United Kingdom had published the Cadbury Code, which was the foundation of today's UK code of corporate governance.

I went to the library—because this was pre-Google—and researched corporate governance and why I would want this job. It sounded like it could be interesting. Long story short, I got the job and was fortunate to determine how it would develop as this was at a very early stage of corporate governance in the United Kingdom. And it turned out that stewardship is a balance of analysis and policy, the combination I was seeking. The company was Hermes Investment Management, which was globally invested but was a major investor in the United Kingdom. Alastair Ross Goobey, CEO of Hermes, had homed in on the idea that predominantly index-tracking investors, like Hermes was on behalf of its clients, should vote actively at shareholder meetings because that's the one mechanism you have to give feedback to companies. At that time, we were focused mostly on voting on the independence of the board. That was over twenty years ago now, and there were only a handful of teams like ours. Of course, the whole area has expanded over time, and we're focused on engagement and voting on a wider range of issues, including directors, executive pay, sustainability, and financial performance.

In 2009, not too long before Alexis, I joined BlackRock. We built the stewardship team to move from a pure focus on proxy voting to, and I would argue, a much more value-added approach of voting supported by direct dialogue with companies. Because roughly ninety percent of the equity assets that BlackRock manages on behalf of clients are in index strategies, our clients are long-term locked-in investors. We see our stewardship efforts as a way to meet our responsibility to clients and to be their voice on issues that are drivers of long-term value and risk. That's exactly where sustainability comes in.

Pagitsas: What is your philosophy as an executive leader, Michelle?

Edkins: One of the things I love about working here is BlackRock's purpose, which has been consistent since the company was founded. While it's been described in a slightly different language, it's always been about helping more and more people experience financial well-being. That is very relatable for me because I grew up in a family that was not well off financially. Financial insecurity is something that I still feel instinctively, even though I am financially much more secure now than I was when I was a kid.

Our mission is to help more people experience financial well-being such that fewer and fewer people experience financial insecurity. That stays with you forever once you've experienced it. It's going to sound trite, but we do make the world a better place. I think for a lot of people here, it's motivating to come into work and have this North Star purpose, supported by a strategy and business plan for how we're going to deliver, this year, this month, this week, especially on the days when so much is going on.

Pagitsas: Alexis, what is your philosophy or approach to moving the sustainability agenda forward?

Rosenblum: I came to the sustainability work from a very different place, from the world of public policy and advocacy. Interestingly, the advocacy skills that I learned through my public policy work have benefited me in my sustainability role. To be successfully adopted, ideas have to proliferate across the organization, and they have to build support from a wide range of internal stakeholders. It's not as easy as demanding a change or making one good argument with one senior or powerful person. It's building up momentum with lots of people gradually over time. It takes a lot of patience and coalition building to do that.

One thing that is so challenging about sustainability work is that you're often on the leading edge, and you have to be patient for others to be ready for the idea. When you're a sustainability professional, you're bringing new ideas into the company and inserting them into the traditional way of thinking. That doesn't always get buy-in overnight. Sometimes we have an idea of the right answer, but it's not necessarily the right time for that answer to be heard or appreciated.

So, it is really about keeping our mission straight and keeping it all in perspective that we're going to have a lot of ideas about different ways to achieve something. But it doesn't mean that just because we think it's the right idea within our smaller team, it's necessarily going to lead to change overnight. It's having that patience to plant the seeds and let them blossom over time. We've seen it so many times over. Eventually someone says a couple months later, "Of course, that's the answer. Now let's run forward."

Pagitsas: You have to be sensitive to the current moment, business situation, and pressures that an organization and individual partners are under, right?

Rosenblum: Exactly.

Pagitsas: Let's think about five years to ten years from now. What do you anticipate or hope will be the state of sustainability and business integration?

Rosenblum: What I anticipate is the same as what I hope, which is that it's all going to be one thing. I don't think there will be sustainability people over here and businesspeople over there. When I took on this role, I had a realization that there is a different set of fundamentals about ESG and sustainability that I didn't learn from my traditional financial training. ESG wasn't really in the curriculum when I went through the CFA Institute charterholder program, for example. Now I'm working with the CFA Institute to incorporate ESG in a new ESG standard they have developed. I think that will change over time, where the traditional educational forums such as business schools increasingly realize that the sustainability fundamentals are critical to their future and the future of their profession.

We see that even with our organization when we talk about ESG integration, ESG is just becoming part of the traditional financial investment decision-making process. It's not a different thing anymore. I think we'll continue to see that grow over time.

Pagitsas: Michelle, looking into the future, what do you hope and expect will be the same or different for ESG and its relation to business?

Edkins: I would hope that real momentum builds around the need for climate action and not just in terms of the big headline things that government needs to do or that private sector needs to do, but also the pretty significant changes that individuals as consumers and citizens have to make because it has to be a collective effort.

It could be just my bias because of where I sit, but we are so focused on what companies can do we sometimes lose sight of the micro changes that are needed. That said, I expect significant change at the consumer level in the next couple of years. There still needs to be a huge amount of awareness-raising of the tangible things that individuals can do. And there is more to understand about the tangible things that companies can do, and the big picture things that government needs to do to effect change. But right now, especially after COP26, there is still too much focus on the big gesture.

In terms of both hope and expectation, we're looking to boards of directors to step up on many fronts. I could have said this every year since I joined this field a couple of decades ago, but the role of directors is so important in ensuring proper oversight structures and minimal missteps. The board needs to provide strong support for executive leaders as they take some very difficult decisions that will put them out of step with certain shareholders, employees, and so on. The boards need to be equipped to counsel and oversee management in those decisions. They need to have the bravery to provide management with cover if a company gets criticized.

Lastly, they need to have patience and persistence because these are long-term shifts that are taking place. It is about having the courage of your convictions that you're on the right course, you're going to keep to the course, and you'll just bear the flack as it comes as long as we are convinced that this is the right thing for the company and its stakeholders, including its investors.

I

Index

Made in the USA
Coppell, TX
19 January 2023

11340007R00197